Advisory Committee on the
Microbiological Safety of Food

Report on Vacuum Packaging and Associated Processes

*Advises the Government
on the Microbiological Safety of Food*

London:HMSO

CONTENTS

ADVISORY COMMITTEE ON THE
MICROBIOLOGICAL SAFETY OF FOOD

The Hon. Nicholas Soames, MP
Parliamentary Secretary
Ministry of Agriculture, Fisheries and Food
Whitehall Place (West Block)
London SWlA 2HH

Secretariat to Advisory Committee
Room 627 Eileen House
80-94 Newington Causeway
London, SE1 6EF
Telephone: 071 972 2924/2985
Fax: 071 972 2892

10 September 1992

Dear Minister,

ACMSF REPORT ON VACUUM PACKAGING AND ASSOCIATED PROCESSES

I have pleasure in enclosing my Committee's Report on its review of the potential hazards of vacuum packaging and associated processes, such as sous vide, with particular reference to the potential risk from *Clostridium botulinum*.

The Report concentrates on the risk of growth and toxin production by psychrotrophic strains of *C. botulinum* in chilled foods, and in particular in those foods with an extended shelf-life. I believe the Report comprises a comprehensive review of the available data and gives practical recommendations for ensuring the continued safety of chilled foods. I do hope that it will be useful to Ministers, and if you agree to publication, to the many people now involved in the production, retailing and consumption of chilled foods.

I hope very much that you will agree to publication of the report which I am sure will be of interest to industry, enforcement authorities and consumers.

I have asked that this letter and the Report be sent to all UK Health and Agriculture Ministers.

PROFESSOR HEATHER M DICK
Chairman

Copies to:

Baroness Cumberlege CBE
Parliamentary Under Secretary of State for Health
Richmond House
79 Whitehall
London SW1A 2NS

The Rt. Hon. David Hunt MBE MP
Secretary of State for Wales
Gwydyr House
Whitehall
London SW1A 2ER

The Rt. Hon. Ian Lang MP
Secretary of State for Scotland
Dover House
Whitehall
London SW1A 2AU

The Rt. Hon. Sir Patrick Mayhew QC MP
Secretary of State for Northern Ireland
Whitehall
London SW1A 2AZ

Advises the Government on the Microbiological Safety of Food

Chairman: Professor Heather M Dick MD, FIBioL, FRCP, FRCPath, FRSE

ADVISORY COMMITTEE ON THE MICROBIOLOGICAL SAFETY OF FOOD

WORKING GROUP ON VACUUM PACKING AND ASSOCIATED PROCESSES

TERMS OF REFERENCE

The Advisory Committee on the Microbiological Safety of Food at its meeting in June 1991 held an initial review of the microbiological aspects of the safety of vacuum packaged and other hermetically sealed foods. The Committee concluded that a careful assessment was needed of the action that might be required to protect the public from any risk of botulism or other dangers. A Working Group was set up to consider the matter in detail and to report back to the Committee. The terms of reference of the Working Group were as follows:

"To prepare a report for the Advisory Committee on the potential hazards of vacuum packing and associated processes such as "Sous Vide", with regard to chilled foods and with particular reference to the risks associated with botulism. Suggested preventative measures should be identified in the report and possible mechanisms for control may also be identified."

The Working Group has now completed its discussions and presents this report to the Advisory Committee.

WORKING GROUP ON VACUUM PACKING AND ASSOCIATED PROCESSES

LIST OF MEMBERS

Chairman

Dr. Michael Stringer,   Director, Food Science Division, Campden Food & Drink Research Association

Members

Mr R. Ackerman   Chairman, Hotel and Catering Training Company

Dr. A. Baird-Parker   Head of Microbiology, Unilever Research Laboratories

Dr. R. Gilbert   Director, Public Health Laboratory Service, Food Hygiene Laboratory

Dr. B. Lund   Formerly Head of Microbiological Food Safety Group, AFRC Institute of Food Research, Norwich

Dr. N. Simmons   Department of Clinical Bacteriology, Guys Hospital

Dr. G. Spriegel   Director of Scientific Services, J Sainsbury Plc

Dame Rachel Waterhouse   Past Chairman, Consumers Association

Assessors

Mr N. Kingcott   Department of Health

Dr. V.King   Department of Health (from 24/3/92)

Dr. R.Mitchell   Ministry of Agriculture, Fisheries and Food

Dr. K.McGrath   Ministry of Agriculture, Fisheries and Food

Secretariat

Mrs D.Linskey   Ministry of Agriculture, Fisheries and Food

Mr P. G. Brophy   Ministry of Agriculture, Fisheries and Food

ACKNOWLEDGEMENTS

The Working Group would like to extend their thanks to the following persons for their assistance with the Group's work:

Dr J Banks	Campden Food & Drink Research Association.
Mr S Bayani	Home Rouxl Ltd
Miss R Blood	British Food Manufacturing Industries Research Association, Leatherhead.
Mr D Clarke	Forte Supplies
Mr J Cole	Sous Vide Advisory Committee
Mr B Day	Campden Food & Drink Research Association
Mrs J Gaze	Campden Food & Drink Research Association.
Miss K Goodburn	Chilled Food Association.
Dr C Hatheway	Centers for Disease Control, Atlanta, Georgia
Prof A Holding	Harper Adams Agricultural College.
Mr J Hutchen	Sous Vide Advisory Committee.
Mr T Miller	Whitbread Plc.
Dr T Roberts	Agricultural & Food Research Council, Institute of Food Research, Reading.
Mr A Roux	Home Rouxl Ltd.
Dr M Sebald	Institut Pasteur, Paris.
Dr M Sheard	Leeds Polytechnic.

The Working Group wrote to the following organisations to inform them about the work of the Group and to seek any information which might aid the Group's work. The Group would like to thank those who responded (denoted by *).

Action Packaging Co Ltd
Association of British Preserved Milk Manufacturers*
Brewers Society*
British Fruit and Vegetable Canners Association*
British Hotel Restaurants and Caterers Association
British Meat Manufacturers' Association*
British Pasta Producers Association
British Poultry Association
British Refrigeration Association
British Retail Consortium*
British Trout Association
British Turkey Federation

Caterers Association of Great Britain

Catering Equipment Manufacturers Association

Chilled Food Association (Food and Drink Federation)*

Co-operative Union Ltd*

Courtaulds Packaging Colodense

Dentech Associates

Dixie Union UK Ltd

DRG Flexible Packaging

Duck Producers Association

Englemann and Buckham Ltd

Flexible Packaging Association

Food Manufacturers Industrial Group

Gatwick Hilton Hotel

Hospital Caterers Association

Hotel Catering and Institutional Management Association*

Institute of Food Science and Technology*

Kempner Ltd*

Lawson Mardon Flexible

Multivac UK Ltd*

National Cold Storage Federation

Pechiney Packaging and Food and General Line Ltd

Planned Packaging Ltd*

Produce Packaging and Marketing Association

Reymar Ltd*

Rhinopac Ltd*

Rhône-Poulenc Ltd*

Refrigeration Industry Board*

Salmon and Trout Association

Sous Vide Advisory Committee*

VACU-IT Packaging (International) Ltd*

Wolf Walsrode*

W R Grace Cryovac Division

INTRODUCTION

1   During the past five years there has been tremendous growth in the chilled food market, with a wide variety of products available to the consumer through retail and catering outlets. In 1991, the chilled food market for ready meals, prepared salads, pizza and pasta was worth £470m compared with only £194m in 1986 - an increase of over 240% in only five years. (Economist Intelligence Unit 1992).

2   The rapid development of new products and the market growth has been influenced in a major way by the large UK retailers with efficient chilled distribution systems and extensive chilled cabinet capacity and the ability to cope effectively with the turnover of short shelf-life products.

3.   The issue of primary concern with all such foods is that of microbiological safety. Current concerns with chilled foods include:

   (a)   Microorganisms - particularly the psychrotrophs which are capable of growing well at low temperatures;

   (b)   Shelf-life - with the increasing use of combinations of preservative factors at sub-lethal or sub-inhibitory levels to maintain the stability of foods, the accurate establishment of a safe shelf-life can be complex;

   (c)   Pasteurisation processes - there is a growing trend towards the use of lower heat treatments (pasteurisation) alone or in combination with other preservation methods in an attempt to retain optimal product characteristics;

   (d)   Temperature control - requirement for control and monitoring of temperature throughout the chill chain.

4.   "Chilled foods" can be defined as perishable foods which, to extend the time during which they remain wholesome, are kept within specified ranges of temperature above -1°C and below +8°C. (IFST, 1990). This report gives particular consideration to prepared chilled foods because of the rapid development in this sector of the food chain and particular concern over the use of extended shelf-lives. Such foods are now required by law to be kept at or below 8°C (5°C from 1 April 1993 for certain foods) throughout the preparation, distribution, retail and catering chain. The Group felt that advice on the hazard posed by the growth and toxin production by psychrotrophic strains of *Clostridium botulinum* in chilled foods should be given bearing in mind that these foods may be exposed to temperatures up to 10°C. In addition it was felt that by considering temperatures of 10°C and below, the hazard presented by the growth and toxin production by non-psychrotrophic, proteolytic strains of *C. botulinum* need not be considered as these strains do not grow below 10°C.

5.   It had been generally considered that chilled foods would remain microbiologically safe if kept in the above temperature range. However, the realisation that some pathogenic microorganisms can grow at chill temperatures challenged that belief. It is now well established that such microorganisms may represent a significant food safety risk and that conditions for their growth and survival must be considered by the food industry.

6.   Because of the reliance on temperature control as the primary or only preservation technique for chilled foods, maintenance of proper temperature control is a critical factor. Temperatures above 8°C may allow most foodborne pathogens to grow. These temperatures may also allow normal spoilage organisms to grow and make the product unacceptable, which in turn would normally result in rejection by the consumer. Many of the new packaging and processing techniques are designed to maintain sensory and microbiological quality for extended times. These techniques may remove or inhibit spoilage microorganisms and allow surviving or contaminating pathogens to grow in the absence of competition. This is particularly true of anaerobic packaging techniques, such as vacuum packaging which restrict the growth of spoilage organisms whilst creating conditions that permit the growth of anaerobic microorganisms such as *C. botulinum*, the psychrotrophic strains of which can grow and produce toxin at temperatures as low as 3.3°C.

SUMMARY

1.   The Group has reviewed the current situation with regard to foodborne botulism and has examined the international evidence of incidents related to extended-life chilled food.

2.   The Group has considered the available data on factors controlling the survival and growth of psychrotrophic *C. botulinum* in foods, and has also reviewed other pathogens of potential concern. For those chilled foods which receive a heat treatment during their production, the Group has recommended time and temperature combinations for the inactivation of spores of psychrotrophic *C. botulinum*. Where the recommended heat treatment is not applied, or a lower heat treatment used, combinations of preservative factors such as pH, water activity (a_w), salt, storage temperature and shelf-life should be established to prevent the growth of psychrotrophic *C. botulinum*. These factors need to be considered for all prepared chilled foods which are assigned a shelf-life of more than 10 days. Chilled foods with a shorter shelf-life should, if kept at chill temperatures, present a minimal risk of the growth and toxin production by *C. botulinum*.

3.   The Group has categorised chilled foods according to the potential for growth of psychrotrophic *C. botulinum* and identified those food categories where the hazard from the growth and toxin production by psychrotrophic strains of *C. botulinum* should be addressed by food manufacturers, caterers, retailers and enforcement authorities.

4.   The Group has considered the information currently available on pasteurisation processes used in chilled food production, on vacuum packaging and on modified atmosphere packaging. It is clear that there is much scientific and technical information on these processes in the public domain. The Group has also explored the current commercial use of these processes and the equipment involved. The Group believes there is a need for a comprehensive code of practice for vacuum packaged and similarly processed chilled foods which should cover all the issues raised by this report plus general hygiene practices. This code should be made available to industry as soon as possible and guidance on its use should be produced for enforcement authorities.

5.   In considering vacuum and modified atmosphere packaged foods the Group has reviewed the regulatory controls currently in place and has looked at possible additional controls such as licensing. The Group believes that registration of food premises should be used by enforcement authorities to identify businesses using processes of concern and that the risk assessment carried out to classify food premises for food hygiene inspections should enable enforcement authorities to target their resources most effectively in these areas.

6.   The Group believes that there is a need to inform consumers of correct handling practices with regard to vacuum and modified atmosphere packaged chilled foods. The Group believes that it is essential that vacuum and modified atmosphere packaged chilled foods should be labelled clearly to ensure that consumers keep them at chill temperatures and that they are not consumed after the "use by" date.

7.   The use of Hazard Analysis and Critical Control Point (HACCP) systems by industry to control their processes should continue to be actively promoted. The Group remains concerned about the lack of knowledge about the hazard of *C. botulinum* among small food processors and consumers. The code of practice recommended by the Group should cover this hazard and promote the use of HACCP principles.

8.   The Group has considered the available data on the factors affecting the survival and growth of psychrotrophic strains of *C. botulinum*. It considers that the information on this and other pathogens which will be available to industry through the MAFF funded Predictive Microbiology Programme and other projects should provide industry with the information necessary to ensure the safe production and supply of chilled foods with an extended shelf-life.

CHAPTER 1

CLOSTRIDIUM BOTULINUM - EPIDEMIOLOGICAL INFORMATION

1.1.   Botulism is a severe illness which is caused by a neurotoxin produced by an anaerobic spore-forming bacterium *C. botulinum*. The initial symptoms of the illness may include nausea, dry mouth and impaired vision with gastrointestinal symptoms of vomiting, slight diarrhoea followed by constipation and abdominal distension and pain. Development of the illness leads to muscle weakness which progresses to flaccid neuromuscular paralysis and affects the respiratory muscles which may result in death if untreated. The clinical symptoms of the illness depend, to some extent, on the immunotype of the toxin and there are reports of mild and atypical forms of botulism. In most countries the illness is rare compared with that caused by other foodborne pathogens. (Further information on laboratory diagnosis can be found in Appendix A, page 52).

1.2.   There are three forms of botulism recognised today; foodborne botulism where toxin is ingested, infant botulism where *C. botulinum* colonises the gastrointestinal tract after ingestion and produces toxin in situ, and wound botulism - a rare illness where *C. botulinum* grows and produces toxin in situ. More than 1000 cases of infant botulism have been reported to date, mainly from the USA where it is now the most common form of botulism (see Table 1, page 11) with a case fatality ratio of less than 4%. Only three cases have been reported in the UK. The classic illness and the most common in the UK is foodborne botulism which is an intoxication and therefore not an infection.

Botulism Worldwide

1.3.   Foodborne botulism has been reported from many countries throughout the world (Table 2, page 11). Type A outbreaks predominate in China, the USA and Argentina, type B outbreaks are most common in Central and Southern Europe and type E outbreaks in Japan, Canada, Iran and Scandinavia.

1.4.   Table 3 (page 12) summarises the food vehicles involved in outbreaks of botulism in various countries. In the USA many outbreaks have been associated with defectively canned or bottled, home produced, low-acid vegetables, while in China they are largely associated with fermented bean curd products. Botulism outbreaks in Europe have mainly occurred in Poland, France, and Germany and involved cured or smoked meats. Table 4 (page 12) shows more detailed data from France where most cases of botulism are due to home processed ham either smoked or salted. Outbreaks in Italy and Spain have been caused mainly by home-preserved vegetables. In all countries, marine products are associated with outbreaks of type E botulism. In Norway the majority of 16 cases between 1975-1991 (WHO Newsletter 1992) were caused by 'rakfisk' a traditional fermented raw fish. Botulism caused by type E is rare in the rest of Europe.

Botulism in the United Kingdom

1.5.   The first reported outbreak of botulism in the UK was in 1922 at Loch Maree in Scotland (Leighton, 1923) when eight people died after eating sandwiches made from wild duck paste. This and subsequent incidents are line listed in Table 5, (page 13). Between 1947 and 1989 only five incidents have been reported. These were associated with macaroni cheese in 1947 (Department of Health, 1948), pickled fish from Mauritius in 1955 (MacKay-Scollay, 1958), canned salmon from the USA in 1978 (Ball et al, 1979), a supposedly shelf-stable prepackaged Kosher airline meal in 1987 (Colebatch et al, 1989) and hazelnut yoghurt in 1989 (Critchley et al, 1989; O'Mahony et al, 1990). The last incident was the largest outbreak of foodborne botulism reported in the UK to date.

1.6.   Home curing of bacon and ham was extensively undertaken in the UK in 1939-1954 with no known related problems of botulism. Home canning of produce has not been practised in the UK since the 1940s and commercial canners have had high standards and control systems. This, together with a lack of tradition for eating raw fish, may account to some extent for the rare occurence of botulism outbreaks in the UK.

Botulism - a changing epidemiology?

1.7.   Between 1976 and 1984, 124 outbreaks of foodborne botulism were identified in the USA involving some 308 persons. Evaluation of the data by MacDonald et al, (1986) implicated both new food vehicles and trends in botulism. Although only 4% of the outbreaks were linked to food served in restaurants, they represented

42% of the cases. Several newly identified vehicles for foodborne botulism were also reported including sautéed onions, baked potatoes, potato salad, pot pies, beef stew and turkey loaf. The foods implicated in these cases were held at 20-50°C overnight. Similarly, in Canada, chopped garlic in soyabean oil implicated in a large restaurant associated outbreak in 1985 had been held at ambient temperatures for several months (Louis et al, 1988). In Taiwan, an outbreak was linked to commercially preserved peanuts (Chou et al, 1988) while the recent outbreak in UK was associated with hazelnut purée used in yoghurt production.

1.8.   Vacuum and modified atmosphere packaging have been in retail use since the early 1960s and the early 1980s respectively. The safety records of vacuum and modified atmosphere packaging are remarkable with only 7 outbreaks (between 1960-1991) recorded worldwide of botulism (5 confirmed and 2 possible/unconfirmed) being associated with these processes (Table 6, page 13). The outbreaks appear to have been a consequence of gross temperature abuse of the product during storage. None of these incidents occurred in the UK.

1.9.   Changes in food processing and preservation such as extended life cook-chill, sous vide catering, vacuum packaging, reductions in nitrite levels in cured meats, and the tendency to use lower cooking temperatures to reduce loss of yield and improve colour and texture of products, all have the potential to cause problems in the future. Use of heat treatments, pH, A_w and preservatives as controlling factors to prevent growth and toxin production by psychrotrophic strains of *C. botulinum* are discussed in Chapter 2 Section 1.

Conclusions and Recommendations

1.10.   In those European countries where the incidence of botulism is higher than in the UK, home preservation of foods such as home curing of meat and bottling of vegetables, is frequently implicated. (C1)

1.11.   Of those few reported incidents of botulism associated with vacuum packaged foods it would appear that different types of smoked fish are the most commonly implicated foods. (C2)

1.12.   Given the changes taking place in food production technology, the Group recommends that food manufacturers critically assess all new food processing procedures to ensure elimination of the risk of botulism. (R1)

1.13.   The Group recommends that home preservation methods such as home canning or bottling of low acid products such as vegetables and meats, and now home vacuum packaging (except for frozen products), should not be encouraged, given the potential risk of botulism. Techniques such as freezing should be advocated for home use. (R2)

TABLE 1. Yearly incidence of botulism in the United States 1973 -1990

YEAR	FOODBORNE	INFANT	WOUND	UNDETERMINED	TOTAL
1973	31	0	1	0	32
1974	31	0	5	0	36
1975	19	1	0	0	20
1976	40	14	3	0	57
1977	86	43	0	0	129
1978	58	39	0	14	111
1979	8	25	3	5	41
1980	18	66	2	1	87
1981	22	71	5	1	99
1982	30	60	1	1	92
1983	43	79	0	3	125
1984	19	99	3	3	124
1985	32	71	1	1	105
1986	24	88	3	0	115
1987	20	86	3	0	109
1988	45	78	2	1	126
1989	23	74	4	0	101
1990	22	74	4	0	100
TOTAL	571	968	40	30	1609

Data compiled and provided by Dr C Hatheway, Centers for Disease Control, Atlanta, Georgia. (Personal Communication)

TABLE 2. Recorded outbreaks and types involved in foodborne botulism

Country	Period	Outbreaks	Cases Total	Fatal[a]	Outbreaks with type identified	A	Type(%) B	E	Other[b]
Belgium	1982-84	7	16	1(6)	7	0	57	14	29(B+C)
W. Germany	1971-82	-	499	33(7)	-	-	>90	-	-
France	1978-84	115	217	6(3)	114	0	96	3	1(AB)
Spain	1969-83	30	93	6(6)	30	0	100	0	0
Italy	1979-83	-	239	-	15	20	60	7	13(A+B)
Czech	1979-84	17	20	0	6	17	83	0	0
Hungary	1961-85	59	109	2(2)	-	-	-	-	-
Poland	1979-83	-	2390	45(2)	1279[c]	2	97	2	0
Denmark	1951-84	11	33	9(27)	9	11	0	78	11(F)
Norway	1962-78	15	36	3(8)	15	0	47	47	7(F)
USA	1971-85	210	485	55(11)	193	63	23	14	0
Alaska	1971-85	35	80	6(8)	33	15	15	70	0
Canada	1971-85	67	165	23(14)	64	3	9	88	0
USSR	1958-64	95	328	95(29)	45	33	38	29	0
Japan	1951-84	96	478	109(23)	96	2	2	96	0
China	1958-83	986	4377	548(13)	733	93	5	1	<1(A+B)
Iran	1972-74	63	170	18(11)	63	0	3	97	0
Argentina	1970-80	20	81	30(37)	14	100	0	0	0

([a]) In brackets % of Fatal Cases

([b]) In parentheses, type(s) identified

([c]) Cases not outbreaks

Adapted from: Hauschild, A. H. W. (1989). *Clostridium botulinum*. In: Foodborne Bacterial Pathogens, ed. M P Doyle, pp. 111-189. New York Marcel Dekker Inc.

TABLE 3. Foods involved in outbreaks of botulism

Country[a]	Outbreaks with food identified	Food (%)				Source	
		Meats	Fish	Fruit and vegetables	Other	Home	Commerce[b]
Belgium	5	60	20	20	0	60	40
W. Germany	-	>75	-	-	-	-	-
France	83	86	5	7	2	88	12
Spain	24	42	0	58	0	92	8
Italy	13	8	8	77	8	-	-
Czech	14	72	7	14	7	100	0
Hungary	58	67	0	3	29	100	0
Poland	2434[d]	87	11	2	0	68	32
Denmark	11	27	73	0	0	-	-
Sweden	6	17	83	0	0	83	17
USA	183	14	17	60	9	90	10
Alaska	33	45[c]	55	0	0	100	0
Canada	63	70[c]	22	8	0	97	3
USSR	83	17	67	16	0	97	3
Japan	94	0	99	1	0	98	2
China	958	10	0	86	4	-	-
Iran	63	3	97	0	0	-	-
Argentina	30	3	10	73	13	77	23

([a]) - or State/Region

([b]) - Includes foods served in restaurants, produced in family-type operations, imported, or abused in the home.

([c]) - Mostly from marine mammals (seal, whale).

([d]) - Cases, not outbreaks

From: Hauschild, A. H. W. 1989. *Clostridium botulinum*. In: Foodborne Bacterial pathogens, ed M P Doyle, pp. 111-189. New York Marcel Dekker Inc.

TABLE 4. Foodborne botulism in France

Year	Outbreaks[a]	Number of cases[b]	Food[c] vehicle	C. botulinum type[c]
1985	11	18	ham (5), pâté(2)	B(9)
1986	12	18	ham (4)	B (10)
1987	20	34	ham (9), pâté (3) , vegetable conserve (1)	B (20)
1988	15	2	ham (5)	B(8), E(1)
1989	18	20	ham (5), pâté (1)	B (12)
1990	28	49	ham (5), vegetable conserve (1)	B (16), A (1)

([a]) In France the term "outbreak" may be used for a single case of botulism

([b]) Cases may be defined as a result of clinical diagnosis + or - biological confirmation of detection and identification of toxin in patient's serum or suspect food.

([c]) Where known.

Table constructed from epidemiological data published in B. E. H. No. 27/1991 "Botulism in 1989 and in 1990 in France", A Pelletier, B Hubert, M Sebald, and from personal communications from B Hubert and M Sebald.

TABLE 5. Foodborne botulism in the UK

Year	Number of cases	Number of deaths	Food Vehicle	*C. botulinum* type
1922	8	8	Duck paste	A
1932	2	1	Rabbit and pigeon broth	?
1934	1	0	Jugged hare	
1935	5?	4?	Vegetarian nut brawn	A
1935	1	1	Mincedmeat pie	B
1947	5	1	Macaroni cheese	
1955	2	0	Pickled fish from Mauritius	A
1978	4	2	Canned salmon from US	E
1987	1	0	Kosher airline meal	A
1989	27	1	Hazelnut yoghurt	B

From: Gilbert, R .J, Rodhouse, J. C., and Haugh, C. A. 1990. Anaerobes and food poisoning. In: Clinical and Molecular Aspects of Anaerobes; VI Biennial Anaerobe Discussion Group, International Symposium, held at the University of Cambridge, 20-22 July 1989. Published by: Wrightson Biomedical Publications Ltd, Petersfield, England.

TABLE 6. Reported incidents of botulism associated with vacuum packaged (VP) and modified atmosphere packaged (MAP) foods.

Year	Country	Food	Number of cases	Number of deaths	*C. botulinum* type
1960	USA	VP smoked ciscoes	3	2	E
1963	USA	VP smoked whitefish chubs	17*	5	E
1965[+]	USA	VP luncheon meat (?)	3	0	?
1970	Germany	VP smoked trout	3	?	E
1984	Japan	VP karashi renkon[a]	36	11	A
1987[+]	USA	MAP shredded cabbage (?)	4	?	A
1991	Sweden	VPsmoked salmon	2	0	E

*   16 cases and 6 deaths reported

[+]   unconfirmed incident

[a]   deep-fried mustard-stuffed lotus root

CHAPTER 2

SECTION 1 : PSYCHROTROPHIC *CLOSTRIDIUM BOTULINUM*

MICROBIOLOGY AND CONTROL IN FOODS

Introduction

2.1. This chapter gives information on the characteristics of psychrotrophic *C. botulinum* and how knowledge of these can be used to prevent the growth or survival of the organism in foods.

2.2. The Government endorsed the Richmond Committee's Report on the Microbiological Safety of Food, Part 1, recommendation that all food processing procedures should be designed on Hazard Analysis and Critical Control Point (HACCP) principles (The Microbiological Safety of Food, Part 1, HMSO, 1990). The HACCP system is a systematic way of analysing the potential hazards in a food operation, identifying the points in the operation where the hazards may occur, and where controls over those that are important to consumer safety can be achieved. These are known as the critical control points (CCPs). The CCPs are regularly and routinely monitored and remedial action, specified in advance, is taken if conditions at any CCP are not within safe and specified limits.

2.3. A potential hazard specifically in relation to *C. botulinum* is the consumption of chilled foods in which the growth and toxin production by psychrotrophic strains of *C. botulinum* may have occurred before the food is perceived to be spoiled. The foods most at risk are those in which the spoilage microflora are eliminated or inhibited, whilst psychrotrophic *C. botulinum* may survive and grow.

Description

2.4. *C. botulinum* is ubiquitous in the soil, in salt and fresh water sediments and in the gastrointestinal tracts of animals and fish. Seven distinct types (A to G) are now recognised and these are differentiated by the serological specificity of their toxins. Types A, B and E account for nearly all human cases of foodborne botulism while type F has seldom been implicated. Type E is often associated with aquatic environments.

2.5. The seven types of botulism are further classified into four subgroups based on physiological differences, with Groups I and II associated with foodborne botulism. The first Group contains type A and proteolytic types B and F while the second Group contains type E and non-proteolytic types B and F. Physiological differences between the Groups are shown in Table 8, (page 19).

2.6. The Group II, psychrotrophic, strains of *C. botulinum* are non-proteolytic, have a growth range between 3.3-45°C and so will multiply and produce neurotoxin at chill temperatures, albeit slowly. Their spores can be destroyed in food by an appropriate pasteurisation process. (See Figure 1, page 20)

Occurrence

2.7. Given the widespread occurrence in the environment of the organism it is not surprising that psychrotrophic *C. botulinum* has been found in many foods and that it is not possible to be certain that an unprocessed food will not contain spores of psychrotrophic *C. botulinum*. All chilled food manufacturers must therefore address the control of psychrotrophic *C. botulinum* through composition and formulation coupled with good manufacturing practice. The only exceptions to this are foods that have been sterilised and then aseptically packaged or have received a sporicidal heat treatment within a hermetically sealed container.

Growth at low temperature

2.8. Figure 1, page 20, illustrates the reported ability of psychrotrophic *C. botulinum* to grow at low temperatures. It can be seen that an incubation period of several weeks is generally needed for toxin formation at the lower end of the temperature range. The lowest established temperature limit for growth and toxin production by strains of psychrotrophic *C. botulinum* is 3.3°C. The upper limit being in the range 40-45°C.

2.9. UK retail and domestic chill cabinets and refrigerators are not designed to achieve the lower limit for growth of psychrotrophic *C. botulinum*. Current retail equipment is designed to meet the 5°C statutory controls

14

which come into force in 1993 (See section on Chill Temperature Controls, Chapter 4 page 36).

2.10.   As shown in Figure 1 (page 20), growth and/or toxin production take longer at lower temperatures. The Group recommends that in order to build in control factors to prevent the growth of psychrotrophic C. botulinum in chilled foods where the organism has not been eliminated or is not sufficiently controlled by some other factor, food manufacturers, caterers and retailers should take account of the time taken for this organism to grow and/ or produce toxin at the actual temperatures the food is expected to encounter throughout its shelf-life. This must take into account storage, transport, distribution, retail, catering and domestic stages as appropriate. **(R3)**

2.11.   It is recognised that certain bulk vacuum packaged foods are stored for a period under strictly controlled conditions at temperatures between -2°C and 0°C prior to final preparation and repackaging for sale. These temperatures are such as to prevent the growth and toxin production by psychrotrophic strains of C. botulinum. However, the Group believes that because it is currently not realistic to maintain temperatures of 3°C or less consistently throughout all parts of the chill chain, it is not acceptable to rely on chill temperatures as the sole method of preventing the growth of psychrotrophic strains of C. botulinum in chilled foods with an assigned shelf-life of greater than 10 days. **(C3)**. The Group recommends that in addition to chill temperatures of less than 10°C (statutory chill temperature controls require a maximum temperature of 8°C where applicable), prepared chilled foods with an assigned shelf-life of more than 10 days should contain one or more controlling factors at levels to prevent the growth and toxin production by strains of psychrotrophic C. botulinum. **(R4)**

2.12.   Growth and toxin production by psychrotrophic C. botulinum have been studied under a range of conditions. The Group has considered the available data on growth of psychrotrophic strains of C. botulinum in foods. The only evidence for growth and toxin production by psychrotrophic C. botulinum below 10°C in less than 10 days are in atypical experimentation conditions which are extremely unlikely to be found in foods. It is the Group's opinion that the risk of toxin production by psychrotrophic C. botulinum in chilled foods stored below 10°C with an assigned shelf-life of less than 10 days is minimal. The controls which are recommended by the Group for the prevention of growth and toxin production by psychrotrophic C. botulinum in chilled foods should be taken to apply to foods intended to have an assigned shelf-life of greater than 10 days. The code of practice recommended by the Group (4.66, page 38) should contain detailed information on the growth rates of psychrotrophic C. botulinum in foods.

2.13.   The Group cannot recommend shelf-life allocations to foods as each shelf-life is dependent on the controlling factors affecting the food and is therefore product dependent. The Group recommends to food manufacturers, caterers and retailers that, as part of good manufacturing practice, sound technical evidence should be available in order to demonstrate to enforcement authorities that an assigned shelf-life is appropriate to ensure the microbiological safety of food (R5). Information on the evaluation of shelf-lives for chilled foods is available to industry (CFDRA Technical Manual No. 28, 1991).

Heat resistance

2.14.   Spores of psychrotrophic C. botulinum are significantly less heat resistant than non-psychrotrophic strains with D(100°C) values generally reported as <0.1 minutes compared with D(100°C) values in the region of 25 minutes. With D(82°C) of 1.5-32 minutes, psychrotrophic type B spores have the highest heat resistance among the psychrotrophic C. botulinum group. The highest D(82°C) value recorded for types E & F is 6.6 minutes in tuna, probably due to protection by proteins and fats. This relative heat susceptibility makes heat treatment of some minimally processed foods a potential controlling factor.

2.15.   D values reported in the literature for these organisms vary. Factors such as choice of strain, type of substrate and the provision of lysozyme in the recovery medium are often responsible for this variability. (Hauschild, 1989).

2.16.   However, research carried out at the Campden Food and Drink Association (CFDRA) for MAFF indicates a D(90°C) value in cod of 1.10 minutes with a Z value of 9 (Gaze & Brown, 1991). This work was carried out using spores of psychrotrophic C. botulinum Type B in cod.

2.17.   These values can be used to calculate the time/temperature combination needed to destroy spores of the organism from foods. Traditionally this is expressed as a "6D" value, which is the time/temperature combination needed to reduce the numbers of microorganisms present by a factor of 10^6 (The Microbiological Safety of Food, Richmond Committee, Part 1, HMSO, 1990)

2.18. Based on the CFDRA research the 6D value for psychrotrophic *C. botulinum* type B can be calculated as a time/temperature combination of 7 minutes at 90°C. Building in a safety margin to allow for varied heat resistance of strains and behaviour in other foods, the 6D value for psychrotrophic *C. botulinum* is 90°C for 10 minutes. Table 7 below shows the equivalent heat treatments at a range of cooking temperatures. It is not recommended that the range of process temperatures is extrapolated to greater than 10°C above or below the reference temperature of 80°C.

TABLE 7. Equivalent 6D time/temperature combinations for spores of psychrotrophic *C. botulinum* in foods

Temperature (°C)	Time (minutes)
70	1675
75	464
80	129
85	36
90	10

An expanded version of Table 7 showing time/temperature combinations from 70 to 90°C can be found in Appendix B, page 53

2.19. The temperature range of 70-90°C is considered to form the practical boundaries of most cooking regimes. However, if a processor wishes to cook at lower temperatures for the purposes of preservation then a 6D cook should be calculated based on research carried out at these lower temperatures for particular products. This can mean cooking for extensive periods of time, for example over 27 hours at 70°C. The implications of such heat treatments need to be fully understood. Results of recently completed MAFF funded work are available which can help guide industry on this. (CFDRA, Technical Manual No. 27 Parts I and II (MAFF project 5240), 1992)

2.20. Unlike the heat-stable toxins of other foodborne pathogens, botulinum toxin is heat sensitive and may be destroyed by heating at 80°C for 10 minutes, or boiling temperatures for a few minutes. (Jay, 1986). However, cooking cannot be regarded as a controlling factor because of the possibility that cooking may not be thorough or even, as in a recent case in Hawaii. (Dairy Food and Environmental Sanitation, March 1992). Foods that require reheating before consumption must be regarded similarly. The Group recommends that because of the potential for lack of control over the cooking or reheating of foods, cooking or reheating should not be relied upon to destroy any botulinum toxin present in foods but that other controlling factors should be used in foods susceptible to the growth of psychrotrophic *C. botulinum* in order to prevent the growth and toxin production by psychrotrophic strains of *C. botulinum*. (R6)

Effects of pH

2.21. The minimum pH reported as permitting growth of psychrotrophic *C. botulinum* is 5.0. Reports of growth below this pH minimum are thought to be due to precipitated proteins forming microenvironments of raised pH within the medium. Upper pH limits for *C. botulinum* growth lie in the range 8-9 and are thought to be of little practical consequence in foods.

Water Activity (A_w)

1. Salt

2.22. At ambient temperatures the growth limiting salt concentration is about 5% for psychrotrophic *C. botulinum*. For chilled foods stored below 10°C the growth limiting salt concentration is 3.5% in the aqueous phase in food. Salt is the a_w depressant mainly used in chilled foods. (Hauschild, 1989).

2. Other Solutes

2.23. In media or foods with salt or other solutes as the main a_w depressant, the growth limiting a_w is 0.97-0.98 for psychrotrophic strains. Other solutes such as sugars and organic acids and lipids are of only limited importance in chilled foods. If these other factors are included in the evaluation of food safety then a_w measurements are essential.

Other factors

Oxygen

2.24.   It is now recognised that the growth of *C. botulinum* in foods does not depend upon the total exclusion of oxygen, nor does the inclusion of oxygen as a packaging gas ensure that growth of *C. botulinum* is prevented. Anaerobic conditions may occur in microenvironments in foods which are not vacuum or modified atmosphere packaged. For example in the flesh of fish conditions which are favourable to toxin production can exist in air packaged fish as well as in vacuum or modified atmosphere packaged fish. However, the particular hazard associated with the latter two systems is due largely to the inhibition of spoilage organisms.

Nitrite

2.25.   Inhibition of *C. botulinum* by nitrite in foods depends heavily upon a number of other factors such as acidity and salt content. In addition, there is pressure to reduce nitrite levels in some foods because of the risk of formation of carcinogenic N-nitroso compounds in some situations. Taken together these two limitations mean that the scope for the use of nitrite on its own to control *C. botulinum* is limited.

Competitive Microorganisms

2.26.   The protective effect of competitive microorganisms can be twofold: inhibition of *C. botulinum* and spoilage of the food such that it would be less likely to be consumed. However, competitive microorganisms cannot be relied upon to prevent growth and toxin production by psychrotrophic *C. botulinum*. Indeed, by reducing the available oxygen they may permit the growth of *C. botulinum* where otherwise it would not have taken place.

2.27.   In foods where competitive microorganisms have been destroyed by heat treatment, and/or the pH has been changed and/or oxygen removed as a result of growth of the facultative anaerobic microorganisms the growth of anaerobes such as *C. botulinum* can take place under conditions where growth would otherwise have been inhibited or the food spoiled before becoming toxic.

Nisin

2.28.   Nisin can be used to control the growth of some Clostridia, in particular *C. tyrobutyricum*, in cheese. It is effective against psychrotrophic *C. botulinum* strains at chill temperatures but is not effective against non-psychrotrophic strains. In the UK nisin is not permitted for use in chilled meat, fish or their products (The Preservatives in Food Regulations 1989, SI 1989/533, HMSO) and is rarely used in cheese other than processed cheese.

Sorbate

2.29.   Sorbate is effective against *C. botulinum* at pH values below 5.5. (Lund <u>et al</u>, 1987; Lund <u>et al</u>, 1990). In the UK sorbate is not permitted for use in chilled meat, fish or their products (The Preservatives in Food Regulations 1989, SI 1989/533, HMSO). For this reason its use as the sole controlling factor for psychrotrophic *C. botulinum* in foods is limited.

Spices

2.30.   There have been reports that some spices have an inhibitory effect on *C. botulinum*. In practice these effects may contribute to the overall control of the organism, but spices may also provide a significant source of botulinum spores in the final product. Also, spices can be used to mask off-flavours in a product that might otherwise be organoleptically unacceptable. Spices cannot be relied upon to contribute to the prevention of growth and toxin production by psychrotrophic *C. botulinum*.

17

Combinations of inhibitory factors

2.31.   Most minimally processed chilled foods currently rely on combinations of controlling factors for their safety. It appears that there are three possibilities for establishing the appropriate combinations. These are as follows:

1. Historical. Given the changes in food composition, handling, presentation and storage now taking place it is not safe to assume that merely because some products have not caused botulism in the past they will not do so in future.

2. Empirical. Challenge testing with spores of psychrotrophic *C. botulinum*. However, challenge testing is complex and subject to a number of limitations, not the least of which is the need to repeat the test if any factors are altered (CFDRA, 1987; Recommendations of the US National Advisory Committee On Microbiological Criteria For Foods, 1990).

3. Research on Control of psychrotrophic *C. botulinum*. MAFF is currently funding a Predictive Microbiology Modelling Programme at a number of UK research centres. Psychrotrophic *C. botulinum* is one of a number of organisms being studied. It is anticipated that knowledge of how the various inhibitory factors interact will facilitate the design of intrinsically safe chilled foods. (See Chapter 6, page 44, for information on Predictive Modelling)

Conclusions and Recommendations

2.32.   The Group has reviewed the evidence available on the factors which can be used to prevent the growth and toxin production by psychotrophic *C. botulinum*. The Group recommends that, in addition to chill temperatures which should be maintained throughout the chill chain, the following controlling factors should be used singly or in combination to prevent growth and toxin production by psychrotrophic *C. botulinum* in prepared chilled foods with an assigned shelf-life of more than 10 days:-

- A heat treatment of 90°C for 10 minutes or equivalent lethality,

- a pH of 5 or less throughout the food and throughout all components of complex foods,

- a minimum salt level of 3.5% in the aqueous phase throughout the food and throughout all components of complex foods,

- an a_w of 0.97 or less throughout the food and throughout all components of complex foods,

- a combination of heat and preservative factors which can be shown consistently to prevent growth and toxin production by psychrotrophic *C. botulinum*, ~~at temperatures up to 10°C~~ (R7)

correctia

18

TABLE 8.   A comparison of *C. botulinum* Groups I and II

Properties*	Group	
	I	II
Psychrotrophic	No	Yes
Proteolytic	Yes	No
Toxin types	A,B,F	B,E,F
Inhibitory pH	4.6	5.0
Inhibitory NaCl Concentration (%)	10	5
Minimal water activity	0.94	0.97
Temperature range for growth (°C)	10-48	3.3-45
Decimal reduction time of spores at 100°C (min)	25	<0.1

*Values when the other conditions are at the optimum for growth

Modified from: Hauschild, A. H. W., 1989. *C. botulinum*. In: Foodborne Bactertial Pathogens, ed M P Doyle, pp. 111-189. New York Marcel Dekker Inc.

FIGURE 1.

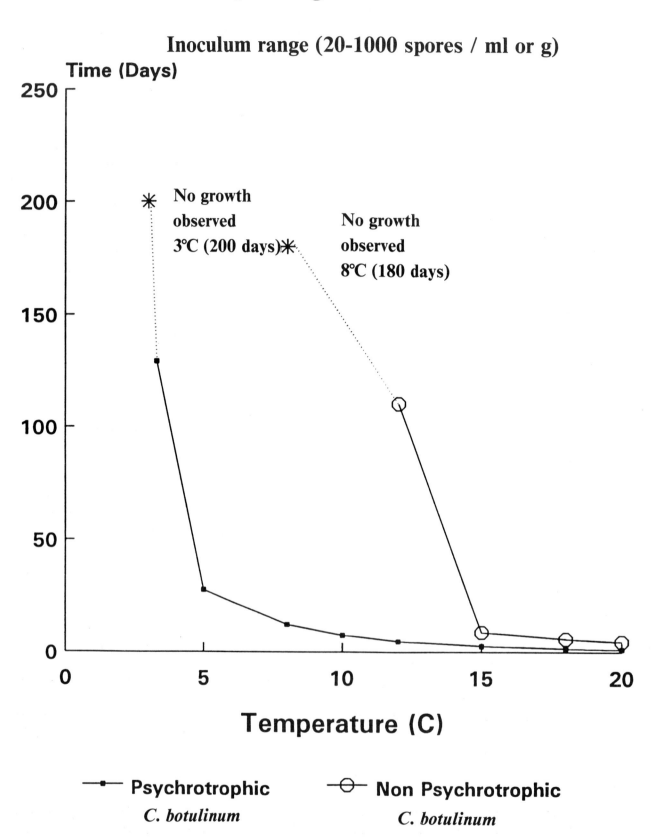

Time to toxin of C. botulinum under Optimum growth conditions.

Inoculum range (20-1000 spores / ml or g)

CHAPTER 2

SECTION 2 : OTHER PATHOGENS OF CONCERN IN CHILLED FOODS

Introduction

2.33.   As well as spores of psychrotrophic strains of *C. botulinum*, a wide range of infectious and toxigenic microorganisms may be present in the raw materials used in the preparation of chilled foods or may contaminate foods during processing. It is necessary to consider the significance of such organisms to the safety of these products and to provide some indication of the methods for their control in chilled foods.

2.34.   Table 9 (page 24) lists the most common foodborne pathogenic microorganisms, their minimum temperatures, pH and a_w for growth, together with their tolerances to air and their resistance to heat. The numerical values in the table are the lowest values recorded for growth under otherwise optimal conditions. The precise limiting values for growth and survival will depend on an interaction of factors. The use of two or more factors in order to control the growth of microorganisms is commonly employed in foodstuffs, including those that are vacuum or modified atmosphere packaged. The D values quoted in Table 9 are for high moisture, neutral pH foods; these values will be higher in foods with a_w of less than 0.97 and will be lower in acid foods (pH values less than 5.0). An example of the interactive effects of three factors on the growth of listeria is shown in Figure 2 (page 25). This also illustrates the type of information which will be available on a number of microorganisms in the Predictive Microbiology Programme. The microorganisms and the interactive factors covered by the Programme are detailed in Table 10 (page 26).

2.35.   The following are brief descriptions of the organisms, their occurrence, illness symptoms and controlling factors. Further information can be obtained from the Report on the Microbiological Safety of Food (Richmond Committee), Parts 1 & 2 (HMSO 1990 and 1991) and the Lancet Review on Foodborne Illness 1991.

Bacillus cereus

2.36.   This spore-forming Gram-positive bacterium is ubiquitous and is a common contaminant of vegetable and dairy products. Illness is self-limiting and can be either of the diarrhoeal or emetic type depending on the strain. Most strains of *Bacillus cereus* do not grow below 10°C. However there are recent reports that some enterotoxigenic strains are capable of growth down to 4°C and are capable of producing toxin at 8°C. *B. cereus* spores are relatively heat resistant and could survive the heat treatment given to some chilled foods during production and/or during reheating. The emetic toxin is extremely resistant to heat so it is important that raw materials used for the preparation of foods are properly stored. Control is by means of selection of the raw materials and limiting the possibility of growth during production eg. rapid chilling of cooked products and proper storage of raw materials and intermediate and final products, combined with control of the packaged product in distribution ie. via time and temperature.

Campylobacter jejuni

2.37.   This Gram-negative non spore-forming bacterium is readily isolated from the intestinal tract of many domestic animal species. Its presence in the environment is probably underestimated because of its particular trait of entering a viable non-culturable form when subjected to stress. It is seldom isolated from processed foods with the exception of raw chicken. It causes a relatively severe illness but recovery is usually complete. It will not grow in chilled foods (minimum growth temperature is 32°C) and currently there is little information as to whether its possible survival is likely to be a concern in chilled foods. The Advisory Committee on the Microbiological Safety of Food is currently studying this pathogen.

Non-Psychrotrophic (Proteolytic) *C. botulinum*

2.38.   This spore-forming Gram-positive anaerobic bacterial species comprises two distinct physiological types; the psychrotrophic type is considered in Chapter 2 Section 1. As with the psychrotrophic strains, these mesophilic, proteolytic strains (that do not grow below 10°C) also cause botulism. The spores of the mesophilic type are very resistant to heat and thus any product capable of supporting the growth of *C. botulinum*, and that will be stored at temperatures above 10°C, must be given a sterilisation treatment known as a "botulinum cook". The control of this organism in chilled foods is principally via control of the distribution and storage temperature so that growth is prevented. Although this organism may grow under poor chill storage it grows very slowly at temperatures below 15°C, and obvious spoilage as a result of other organisms surviving a heat treatment will most likely occur well before this organism has reached levels which may cause botulism. Control is achieved by the use of raw materials with low numbers of sulphite-reducing anaerobic spore-formers, good manufacturing practices (particularly proper chilling). In addition, there are single controlling factors of: 10% salt in the aqueous phase (throughout the food), a_w of 0.94 and pH of less than or equal to 4.5, which can be used to control the growth of non-psychrotrophic *C. botulinum* in foods. If these measures are judged likely to be ineffective then a botulinum challenge study on the product, using mesophilic strains, may be necessary.

Clostridium perfringens

2.39.   This is another mesophilic spore-forming Gram-positive anaerobe that occurs widely in nature. The illness is relatively mild and self-limiting with severe abdominal cramps being a particular characteristic. It is of no concern in vacuum packaged chilled foods in an adequate chill chain as it will not grow below 15°C and grows only very slowly at temperatures below 20°C. For ambient stored vacuum packaged products capable of supporting growth of this organism, sterilisation must be used, the "botulinum cook" effectively destroys this organism.

Escherichia coli

2.40.   This Gram-negative, mesophilic, non spore-forming bacterium, is widely spread in the environment, the main source being the intestinal tract of animals. Some strains cause a wide range of illnesses depending on the toxins produced by individual strains; these vary from severe illnesses caused by the verotoxin producing strains (such as serotype 0157) to relatively mild intestinal illness caused by the enterotoxin producing strains. This organism will not grow at temperatures below 7°C therefore is of little concern in chilled food within an adequate chill chain. Control is further achieved by way of selection of raw materials and the use of good manufacturing practices and distribution conditions that inhibit growth of microorganisms and limit re-contamination and growth of contaminating microorganisms.

Listeria monocytogenes

2.41.   This Gram-positive aerobic non spore-forming bacterium is a general environmental contaminant and traces may be found on virtually all raw agricultural products and many processed ones. It is generally regarded to be of particular concern in chilled foods, because of its ability to grow at temperatures close to freezing and to survive/grow under a wide range of pH and a_w values. Human infection may result in meningitis, septicaemia and infection of the foetus. Infection "in utero" may cause foetal death or the neonate may succumb to generalised infection. Immuno-compromised individuals are also particularly susceptible to listeriosis. Control is achieved by a variety of techniques depending on the particular product; these include heat i.e. heating at 70°C for 2 minutes or an equivalent heat process; the selection and disinfection of raw materials to reduce the number and incidence of contamination; the use of high care areas in the manufacture of foods to reduce the risk of contamination of cooked materials; and limited shelf-life of products in the market place.

Salmonella

2.42.   This genus consists of Gram-negative non spore-forming mesophilic rods widely distributed in nature with man and animals as their primary reservoirs. Salmonella "food poisoning" is caused by ingestion of the organisms in foods. Symptoms usually develop between 12-36 hours after ingestion although shorter times have been reported. Symptoms include nausea, vomiting, diarrhoea, abdominal pain, headaches, moderate fever and muscle weakness. The most common food vehicles of salmonellosis in man are poultry, eggs, meat and meat products, although chocolate and other products have occasionally been implicated. Minimum reported growth temperatures are 5.3°C (*Salmonella heidelberg*) but these are usually higher e.g. 6.2°C for *Salmonella typhimurium*. Therefore growth in properly chilled foods is unlikely to be a problem. Control can be further achieved by selection of raw materials, the use of good manufacturing practice and of distribution conditions that limit recontamination and growth of contaminating organisms.

Staphylococcus aureus

2.43.   This Gram-positive non spore-forming coccus is a widely occurring commensal of warm blooded animals, although almost all food poisoning is a result of contamination by strains of human origin. It causes a self-limiting illness (intoxication) as a result of the production of an emetic toxin in a food. *Staphylococcus aureus* grows very poorly or not at all under chill conditions and does not grow well in the presence of competing microorganisms. Its production of toxin is limited to temperatures above 10°C, while packaged anaerobically toxin is not likely to be produced in significant amounts at temperatures below 15°C. Control in chilled foods is by selection of raw materials which must be prepared/processed and stored properly to prevent growth and toxin formation as the toxin is very stable and particularly resistant to heat. The organism has a similar heat sensitivity to cells of *Listeria monocytogenes*.

Vibrio parahaemolyticus

2.44.   This a Gram-negative mesophilic marine vibrio that is generally associated with fish and shellfish caught in coastal and estuarine waters in warm climates. This organism grows very poorly at chill temperatures, is a moderate halophile (requires 3% NaCl for growth), and control is via choice of raw materials and good manufacturing and distribution practices.

Yersinia enterocolitica

2.45.   This is a psychrotrophic Gram-negative non spore-forming organism that grows down to a temperature of -1°C. The illness it causes is generally a self-limiting enterocolitis but some persons suffer severe sequelae including arthritis, septicaemia and metastatic abscesses. It is widely found in the animal kingdom and of the food animals, pigs are often contaminated with disease-causing strains. Yersinia will grow well in vacuum packaged products, particularly in the absence of competing microflora, and in the absence of any other controlling factors such as pH or lowered a_w. Carbon dioxide combined with good temperature control will prevent growth. Control is generally achieved by the selection of raw materials and control of manufacturing and distribution conditions.

Hepatitis A virus

2.46.   It is currently the view that most viruses causing foodborne illness are of human origin and inclusion of the Hepatitis A virus in the table is illustrative of an example of a particularly heat resistant virus. All enteroviruses will survive well at chill temperatures and be expected to survive, although not to multiply, in most vacuum packaged foods. Control is by way of raw materials and in particular the personal hygiene of operatives handling raw materials and final products prior to packaging.

Conclusions

2.47.   With the exception of anaerobic clostridia, most bacteria which cause foodborne disease grow best in the presence of air and therefore in vacuum and modified atmosphere packaged foods growth will usually be slow. However, risk of growth may be increased as a result of contamination of a sterilised or pasteurised product after thermal processing. This possibility is widely recognised and is the subject of standard procedures in industry where the choice of packaging materials and means of assuring pack integrity, combined with hygienic procedures for handling product post-thermal processing, are designed to prevent re-contamination. Risk of growth may also be increased by abuse of the chill chain. The introduction of statutory temperature controls has lead to a greater awareness of the need to monitor and control temperatures throughout the chill chain (C4).

TABLE 9. Growth/Survival Characteristics of The Most Common Foodborne Disease causing Microorganisms[+]

Microorganisms	Growth[*]			Aerobic/ anaerobic	Heat resistannce (mins)		
	Min temp°C	Min pH	Min a_w		D70°C	D90°C	D121°C
Bacillus cereus	4	4.3	0.91	Facultative	-	10+	-
Campylobacter jejuni	32	4.9	0.99	Microaerobic	0.0001	-	-
C. botulinum (mesophilic)	10	4.6	0.93	Anaerobic	-	-	0.2
C. botulinum (psychrotrophic)	3.3	4.8	0.97	Anaerobic	-	1.5	-
Clostridium perfringens	15	5.0	0.95	Anaerobic	-	-	0.15
Eschericia coli	7	4.4	0.95	Facultative	0.001	-	-
Listeria monocytogenes	0	4.3	0.92	Facultative	0.3	-	-
Salmonella	6	4.0	0.94	Facultative	0.001-0.01	-	-
Staphlococcus aureus	6	4.5	0.86	Facultative	0.1	-	>1.0
	10 for toxin	5.2 for toxin	0.90 for toxin		(cells)		(toxin)
Vibrio parahaemolyticus	5	4.8	0.94	Facultative	0.001	-	-
Yersinia enterocoltica	-1	4.2	0.96	Facultative	0.01	-	-
Hepatitis A virus	Survives Freezing	Survives Mod acid (pH 4.0)	Killed by Drying (most food)	-	-	0.2	-

[*] Survival will usually occur under conditions not allowing growth (survival time will depend on the particular growth limiting condition, composition of the food, and packaging conditions).

[+] under otherwise optimal conditions.

References: 1. Microbiological Safety of Foods (Parts I and II)

2. Foodborne illness - A Lancet Review (1991) Edward Arnold London

3. Microfacts - (1991) Leatherhead Food Research Association

FIGURE 2.

EFFECT OF pH AND SALT CONCENTRATION
ON THE GROWTH OF *Listeria monocytogenes*
AT 10°C

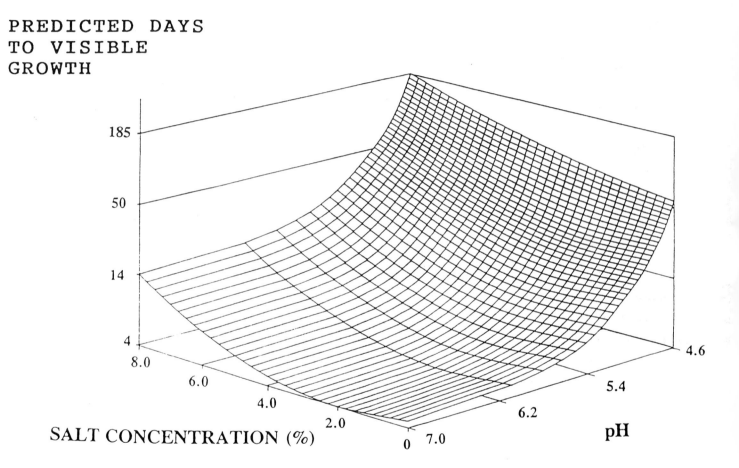

PREDICTED DAYS
TO VISIBLE
GROWTH

SALT CONCENTRATION (%)

pH

(Source: Unilever Research)

TABLE 10.   Matrix showing work under the Predictive Microbiology Programme

THE GROWTH MODELS CURRENTLY AVAILABLE ARE:-

Organism	Temperature	pH	NaCl	Nitrite	Lactic acid
Salmonella	*	*	*	*	
L. monocytogenes	*	*	*	*	
Y. enterocolitica	*	*	*		*
S. aureus	*	*	*		
B. cereus	*	*	*		
B. subtilis	*	*	*		
C. botulinum (psychrotrophic)	*	*	*		
A. hydrophila	*	*	*		

Predictive models for the growth of E. coli and Campylobacter will also be available

CHAPTER 3

CONTROLLING FACTORS OF PSYCHROTROPHIC *CLOSTRIDIUM BOTULINUM* IN CHILLED FOODS

3.1.   Chilled foods for the purposes of this report may be defined as those which require holding at chill temperatures to maintain safety. During manufacture they may be subjected to a heat process stage, but if heat treatment does occur, this maybe insufficient to destroy spores of psychrotrophic *C. botulinum* although heat will often increase the effectiveness of other controlling factors in foods.

3.2.   In chilled foods, psychrotrophic *C. botulinum* will, therefore, often be controlled by inhibition rather than destruction. The controlling factors in a chilled food may act singly or in combination. The safety of any given food will depend upon the number and value of each of the controlling factors acting in combination to exert a combined effect on the organism. Each factor is sometimes known as a hurdle and the use of combinations of factors is commonly known as the "Hurdles Concept".

3.3.   Traditional food preservation techniques, such as salting, smoking and the addition of preservatives such as nitrite, are still in use, although the trend in recent years has been toward reducing levels of these chemicals in foods. As these levels are reduced, the margin of safety afforded is reduced, and the collective effect of the "hurdles" acting on the microorganisms is less. Control must, therefore, rely on other principal inhibiting factors, usually chill storage, and reduced shelf-life to assure product safety.

3.4   Such limitations have, in some cases, motivated industry to develop innovative means of extending shelf-life of traditional products or to design products to have extended shelf-lives. Vacuum or modified atmosphere packaging of foods is usually accompanied by shelf-life extension of the product compared with aerobically packaged equivalents.

3.5.   As chilled foods are not, as a group, sufficiently heat treated to destroy spores of psychrotrophic *C. botulinum*, all chilled foods must potentially be regarded as at risk from this organism. The level of risk will depend upon the presence of the organism, the product formulation and composition, processing technique and form of packaging. These will determine the range and value of inhibitory factors, and consequently the degree of risk. As discussed in Chapter 2, those chilled foods at risk from the growth and toxin production by psychrotrophic *C. botulinum* are those with an assigned shelf-life of more than 10 days.

3.6.   The Group considered that identification of the chilled foods at risk from psychrotrophic *C. botulinum* was best achieved by drawing up a reference table of categories of food which are maintained under chill storage. Some of these are, or may be, vacuum or modified atmosphere packaged. These foods were then categorised on the basis of their composition, processing, and form of packaging and the usual controlling factors noted. A priority for attention classification was then given to each category indicating those considered to present a high, medium, or low risk from growth and toxin production by psychrotrophic strains of *C. botulinum*.

3.7.   The controlling factors used in the assessment are set out in Table 11 (page 28). The categorisation of foods and priority for attention classification rating are detailed in Table 12 (page 29). The information contained in Table 12 was drawn up as a working reference for the Group and it is not intended that it be used independently of the report, or regarded as a definitive list of high risk chilled foods. Table 12 should not be applied outside the scope of this report.

3.8.   Table 12 represents a very broad categorisation of chilled foods. Clearly products within each category may vary according to the manufacturers, retailers or caterers recipe or formulation. This emphasises the need for products to be appraised individually for the *C. botulinum* hazard.

3.9.   The basis for the three categories is as follows:-

Low Priority For Attention

3.10.   Product categories regarded as low priority for attention include those where psychrotrophic strains of *C. botulinum* are either unlikely to occur, or those foods where the product composition results in the presence of a number of controlling factors, some of which may, on their own, be at or above the level required to inhibit growth and toxin production by *C. botulinum*. In certain foods spoilage by aerobic bacteria, although not a reliable controlling factor, will alert consumers to unfit food, before production of toxin would reach unsafe levels. The short shelf-life of other foods (2-3 days), evidence of activity of other microorganisms and storage at chill temperatures mean that *C. botulinum* is unlikely to have the opportunity or time to produce toxin. Examples of food in the low priority for attention category include yoghurt (pH <5.0 - but changes in formulation and added ingredients could move yoghurt into the medium category), dressed salads (pH <4.0), aerobically packaged sandwiches, fresh meat and fresh fish. The foods in this category are regarded as presenting a low risk to the public of the botulinum hazard occurring.

Medium Priority For Attention

3.11.   Products classified by the Group as medium priority for attention include those in which it would not be uncommon to detect the presence of *C. botulinum*, and where no single controlling factor occurs at a level inhibitory to growth. Some products, because of their heterogeneous nature, may provide isolated areas within the product favourable to growth and toxin production by *C. botulinum*. Others may have added ingredients which may contribute spores to the food. Foods in the medium priority for attention category do generally contain more than one controlling factor but the level at which these occur will be product specific. Foods which are included in this category should be regarded as requiring the *C. botulinum* risk to be addressed in their product design. Examples include some meat, fish and vegetable pâtés, and some filled pasta products.

High Priority For Attention

3.12.   Foods are classified by the Group as high priority for attention because abuse of shelf-life or chill temperature conditions could result in growth and toxin production by psychrotrophic strains of *C. botulinum*. The Group <u>recommends</u> that the risk of the *C. botulinum* hazard in these chilled foods must be addressed by manufacturers, caterers, and retailers. Examples include smoked salmon and trout, cuisine sous vide products and modified atmosphere packaged sandwiches (**R8**).

Conclusion

3.13.   The Group's arbitrary assignments of priority for attention to categories of chilled foods are designed to help its work and should not be applied outside the scope of this report. However, they may be used in the preparation of the code of practice recommended in this report to help advise manufacturers, retailers, caterers and enforcement officers of those categories of chilled foods which in the Group's opinion are those most likely to present a risk to the growth and toxin production by psychrotrophic strains of *C. botulinum* (**C 5**).

TABLE 11.   Factors Controlling Growth and Toxin Production by Psychrotrophic *Clostridium botulinum* in Chilled Foods

1.	Water Activity (a_w)
2.	pH
3.	Chill Temperatures
4.	Salt
5.	Nitrite
6.	Shelf-life
7.	Spoilage by Aerobic Bacteria*

* Not essentially a controlling factor but growth of spoilage bacteria may suggest food is unsuitable for consumption.

TABLE 12. Risk assessment of psychotrophic strains of *C. botulinum* in chilled foods.

The information contained in this table is intended as a working reference and it is not intended that it be used independently of this Report

NO.	FOOD CATEGORY	EXAMPLES	USUAL CONTROLLING FACTORS	PRIORITY FOR ATTENTION
1	Raw Animal Products	Fish, Poultry Shellfish, Meat	3,6,7	Low
2	Cooked, uncured, smoked	Meat, Fish, Shellfish Poultry, Frogs legs, Snails, Chicken roll	Air: 3,6,7 MAP: 3,6	Medium High
3	Cooked cured	Meat, Poultry	3, 4, 5, 6	Low
4	Hot smoked	Mackerel/Trout, Shellfish	3, 4, 6	High
5	Cold smoked	Salmon	3, 4, 6	High
6	Uncooked, cured	Bacon	3, 4, 5	Low
7	Uncooked, Cured, Dried	Parma Ham, Bressoala	1, 3*, 4, 5	Low
8	Uncooked, Cured, Fermented	Salami	1, 2, 3*, 4, 5	Low
9	Pâtés:			
	1. Meat, Cooked, Cured	Pork Liver	3, 4, 5, 6	Low
	2. Meat/Fish, Cooked, Uncured, Smoked	Chicken, Mackerel, Trout, Tuna, Salmon	3, 4, 6	Medium
	3. Vegetables, Nuts	-	3, 6	Medium
10	Quiche/Flans	-	3, 6, 7	Low
11	Savoury Mousses	Salmon	3, 6, 7	Low
12	Sandwiches:			
	- Air Packaged	-	3, 6, 7	Low
	- MAP	-	3, 6	High
13	Vac Packaged MAP	Chopped Salads	3, 6	Low
	Salads and vegetables	Chopped Raw Veg	3, 6	Low
		Cooked Veg	3, 6	High
14	Dressed Vegetable or Cereal Salad	Potato, Rice, Bean Salad Coleslaw, Nut Salad	2, 3	Low
15	High Acid Dairy Products pH <5.0	Yoghurt, Fromage Frais Cottage Cheese, Yoghurt Drinks	2, 3**	Low
16	Low Acid Dairy Products pH>5.0	Mousse, Dairy Desserts, Creme Caramel	3, 6, 7	Low
17	Dips:			
	a. Homogenenous	Hummus	2, 3, 6	Low
	b. Heterogenous	Tzatziki	2, 3, 6	Low
18	Soft Cheese (Mould Ripened)	Brie	2, 3, 4	Low
19	Hard Cheese	Cheddar	1, 2, 3**, 4	Low
20	Processed Cheese Spread	-	1, 2, 3**, 4	Low
21	Pasteurised Milk and Flavoured Milk	-	3, 6, 7	Low
22	Dough Products (Raw or Part Baked)	-	3, 6, 7	Low
23	Fresh Chilled Pasta	Cannelloni, Ravioli	Air: 3, 6, 7 MAP: 3, 6	Low Medium
24	Chilled Ready Meals (other than Sous Vide)	-	3, 6, 7	Low
25	Cuisine Sous Vide	-	3, 6	High

* Some products in this category will be ambient stable
** Use of chill temperatures is not required as a controlling factor for *C. botulinum*

Usual Controlling Factors:
1. Water Activity
2. pH
3. Chill Temperatures
4. Salt
5. Nitrite
6. Shelf-life
7. Spoilage by aerobic bacteria (not essentially a controlling factor but growth of spoilage bacteria may suggest food is unsuitable for consumption).

CHAPTER 4

PACKAGING AND MANUFACTURING PROCESSES

Introduction

4.1   There are several processes which result in either vacuum or modified atmosphere packaged chilled foods reaching the catering and retail markets. Vacuum packaging is an excellent packaging format for many products but there are disadvantages and some of these can be overcome by using modified atmosphere packaging. For example, sensitive products can be unduly compressed, slices of cured and cooked meats cling together and are difficult to separate, fresh meat loses its red colour and certain types of cheese tend to appear ''wet'' upon pack opening.

4.2.   This chapter considers these packaging and manufacturing processes and discusses issues relating to heating, chill storage, distribution and good manufacturing practice.

VACUUM PACKAGING

4.3.   Vacuum packaging is an established packaging technique which has been in use for over 30 years. It involves the removal of all or most of the air within a package without deliberate replacement with another gas mixture and is used to extend shelf-life by inhibiting the growth of aerobic spoilage microorganisms and reducing the rate of oxidative deterioration. In order to maintain anaerobic conditions around a food, high O_2 barrier materials are necessary. Although the requisite O_2 barrier for vacuum packaging depends on the type of food packaged, O_2 transmission rates of <15 cm^3 m^{-2} day^{-1} atm^{-1} are generally required. Also, packaging materials with low water vapour transmission rates must be used.

4.4   One of the main traditional uses of vacuum packaging has been for primal cuts of red meats and bacon. More recently vacuum packaging has been used for retail packs of cooked meats, pâtés, fish, and for prepared vegetables. It is only in recent years that vacuum packaging has been used within the catering industry, for use as a food preservation / preparation technique, and also for use as a cooking technique e.g. cuisine sous vide. In addition, vacuum packaging machines are now being sold on the domestic market, and also to small businesses where the knowledge of the potential risks involved with the vacuum packaging process may not be realised.

Vacuum Skin Packaging (VSP)

4.5.   Vacuum skin packaging (VSP) is a packaging technique which was developed to overcome some of the disadvantages of traditional vacuum packaging. The VSP concept relies upon a highly ductile plastic barrier laminate which is gently draped over a food product, thereby moulding itself to the actual contours of the product to form a second skin. The product's natural shape, colour and texture are highlighted and, since no mechanical pressure is applied during the vacuumising process, soft or delicate products are not crushed or deformed. Successes of VSP in the UK marketplace include sliced cooked and cured meats, pâté and fish products (e.g., peppered mackerel). Unlike vacuum packaging, VSP allows pre-sliced meats to be easily separated after pack opening. Since the bottom and top web films are sealed from the edge of the pack to the edge of the product, pack integrity is maximised and juice exudation is limited. VSP saves space in domestic refrigerators compared to modified atmosphere packs and is ideally suited for freezing since the second skin prevents formation of ice crystals on the product surface.

VSP/Modified Atmosphere Packaged 3-web Packaging

4.6.   This is a new development which combines the advantages of both VSP and modified atmosphere packaging primarily for fresh red meat packaging. A standard VSP pack for fresh meat is produced with a permeable top web skin. After this process, the product enters a second sealing station where a third web is used to place a lid on the pack. The space between the top web skin film and the lid is flushed with an O_2 rich gas mixture and, because the top web skin is oxygen permeable, the meat remains bright red in colour. Another application for VSP/modified atmosphere packaged 3-web packaging is seafood.

Vacuum Packaging Equipment

4.7.   There is a wide range of commercially available equipment for undertaking vacuum packaging of foods. The Group recommends that food manufacturers and caterers should ensure that they have a thorough understanding of the operational capabilities of their machines; that the appropriate pouch or tray materials are used; and that the machinery used for establishing a vacuum functions to their required specification **(R 9)**.

4.8.   Pouches must be heat-tolerant, puncture resistant, and provide an effective barrier to oxygen. EC legislation, both current and proposed, must be taken into account. Currently for plastics there is an overall migration limit, and a list of monomers and other starting substances, some with specific migration limits.

4.9.   The Group recommends that all vacuum packaging machine manufacturers should include a section in their instruction handbook which alerts the user to the food safety hazards and the risks from organisms such as *C. botulinum*. These issues should also be explained during equipment demonstrations **(R 10)**.

Cuisine Sous Vide (Sous Vide)

4.10.   "Cuisine sous vide" or vacuum cooking was developed in France in the early 1970s by Georges Pralus who initially was challenged with the task of reducing the shrinkage of foie gras. He developed a vacuum cooking process utilising multi-layered plastic pouches which reduced the shrinkage from 16% to only 5%. The Group considers the terms "cuisine sous vide" and "sous vide" as synonymous.

4.11.   The cuisine sous vide system involves the following stages:

    (i)   Preparation of raw materials and partial cooking if necessary

    (ii)   Food is packaged in heat-stable, air and moisture impermeable plastic pouches.

    (iii)   Vacuum is applied to the pouch - thus removing air, and the pouch is sealed.

    (iv)   The pouches are cooked according to a specific time- temperature pasteurisation treatment.

    (v)   The pouches are then rapidly cooled and packaged into protective cartons.

    (vi)   The products are then stored at or below 3°C.

If the product is to be reheated this can be done in the pouch in moist heat or the contents removed and reheating undertaken by conventional methods.

4.12.   There are a number of potential hazards with the cuisine sous vide process, including: the thermal process used which may not destroy spores of *C. botulinum*, the anaerobic conditions produced within the pouch, the conditions of cooling following pasteurisation, and the lack of an effective chill chain, any of which may permit growth of *C. botulinum*. These potential hazards should be analysed using HACCP principles (see paragraph 2.2). A number of papers have been published describing the application of HACCP to sous vide products. (Adams 1991), (Smith et al, 1990), (Snyder 1991).

4.13.   The heat treatment given to cuisine sous vide products is critical to ensuring their microbiological safety. A MAFF funded study on the heating efficiency of equipment available to caterers in the UK for cooking sous vide products was commissioned in 1990. The study involved trials on eleven ovens using both gas and electricity for power. Each oven was loaded with eight chilled standard sous vide packaged potato slabs evenly spaced on each shelf. The ovens were set on a sous vide cooking programme at 80°C. The study concluded that although performance varied between ovens, none of the sous vide cooking equipment tested heated sous vide packs uniformly when fully loaded and set at an operating temperature of 80°C. (Sheard, 1991)

4.14   The Group is concerned about the lack of uniformity shown by sous vide cooking equipment and recommends that the cooking of sous vide products should not be undertaken unless operators have available the technical expertise to ensure that pasteurisation equipment and operating procedures are adequately designed and tested to give a uniform and known heat load to all containers during each cycle **(R 11)**. As with the canning of low acid foods test results should be available as part of good manufacturing practice to

31

demonstrate the adequacy of procedures to obtain uniform temperatures throughout the load of containers.

4.15   The Group recommends that manufacturers and users of heating equipment used for the production of foods cooked under vacuum ensure that their equipment is capable of achieving the performance they require **(R 12)**.

The Market for Cuisine Sous Vide Products

4.16.   In France, the home of cuisine sous vide, the market is estimated at 2,000 in-house producer users with around 20 manufacturers. Production for retail is aimed at the hypermarkets. The market for cuisine sous vide products has been estimated at 8,000 to 10,000 tonnes per annum. Products vary from traditional dishes, to nouvelle cuisine and to ingredients such as sauces. One French company alone was reported to have sold more than 7 million sous vide dishes, representing 200 different recipes, throughout Europe in 1990. This company has introduced dishes into the UK catering market through an agent. Each product has an assigned shelf-life of 21 days and comes in both multi- and single-portion packs. At the present time in the UK there are no known sous vide products on sale in the retail market. Home production is either for in-house use by, for example, large hotels, or for sale within catering outlets.

Microbiological Status of Sous Vide Products

4.17.   Since there was little published information on the microbiology of cuisine sous vide products marketed in the UK, MAFF commissioned a study in November 1990 to examine the microbiological status of such products at time of purchase and after storage for 21 days at temperatures of 3.3°C and 8°C. Four companies provided products. Three of these companies were producing in the UK and the fourth imported products from France. (Blood, 1991).

4.18.   Duplicate packs of 24 cuisine sous vide products were examined for their microbiological condition as soon as possible after manufacture and after a further 21 days storage at 3.3°C and 8°C. Packs were examined for their aerobic and anaerobic colony counts and for appropriate indicator organisms and pathogens.

4.19   On receipt, all packs had very low total colony counts and no indicator organisms or pathogens were detected. After chill storage the majority of packs were also of very good microbiological quality. However 9 of the 96 stored packs had aerobic colony counts greater than 5×10^4 colony-forming units per gram, seven packs after storage at 8°C and two after storage at 3.3°C. On occasions, considerable variation between packs was observed.

4.20.   Very few indicator organisms/potential pathogens were detected in the stored samples in this study, although *Yersinia enterocolitica* was detected in two packs of one product (25g samples). Strains of Bacillus spp. constituted part of the high colony count flora of five packs, some of which were identified as *B.subtilis*.

4.21.   The survey of cuisine sous vide products commissioned by MAFF shows little cause for concern about the microbiological status of cuisine sous vide products currently available in the UK. However, the survey was limited and consideration should be given to a further survey on a larger scale. **(C 6)**. The Group recommends that the Steering Group on Microbiological Safety of Food consider undertaking surveillance of vacuum and modified atmosphere packaged foods, including sous vide products, focusing on those areas designated a high priority for attention in Chapter 3 of the report **(R 13)**.

The Extent of Sous Vide and Vacuum Packing in England and Wales

4.22.   To enable the Group to have a clearer picture of the current position regarding the extent of the use of cuisine sous vide and vacuum packaging in the market place a questionnaire was sent to all Chief Environmental Health Officers in England and Wales on 30 August 1991. This survey was carried out prior to the introduction of registration of food premises on 1 May 1992.

4.23.   The questionnaire asked for information on the number of sous vide and vacuum packaging operations within a local authority area. The information received is summarised in Table 13, (page 39).

4.24.   Further details were requested from 10 of the Environmental Health Departments who reported sous vide production units in their areas. On investigation it was found that:-

3 had now discontinued use or gone out of business;

2 were demonstration units in catering colleges;

1 supplied curry sauce "boil in bag" frozen product to retail outlets;

1 ran a "hot fill" operation supplying chilled vegetarian products to retail outlets and to airlines, not classified as a sous vide process;

1 produced 30 sous vide meals per day integrated into a conventional catering unit;

1 pub supplied meals (possibly "hot fill" not sous vide) to two others;

1 large manufacturer of sous vide products supplied catering establishments.

4.25.   From the responses received to the questionnaire sent to Environmental Health Departments the Group concluded that only a small number of production units for cuisine sous vide, or similarly produced food, are known to Environmental Health Departments in England and Wales.

4.26.   A majority of local authority areas contain premises vacuum packaging raw meat. Around 10% of areas contain premises vacuum packaging raw fish and vegetables. 44% of areas contain premises vacuum packaging pre-cooked food in factory premises. This could include a range of foods such as smoked fish, cured meats, cooked meats etc. Retail premises vacuum packaging pre-cooked foods occur in over 30% of areas.

4.27.   Several Environmental Health Departments, in submitting their completed questionnaires, expressed concern over the safety of the growing use of vacuum packaging, particularly in retail premises. It is clear that local authorities would welcome further information and guidance on processes of this type.

4.28.   The Group recommends that enforcement authorities use the data which will be available to them under the Food Premises (Registration) Regulations 1991 to identify the types of food premises using vacuum packaging and similar processes and to target food hygiene inspections and other enforcement activities under the Food Safety Act 1990 (R 14).

MODIFIED ATMOSPHERE PACKAGING

4.29.   Modified atmosphere packaging is a food preservation technique whereby the composition of the atmosphere surrounding the food is different from the normal composition of air which is 78.08% nitrogen, 20.96% oxygen, 0.03% carbon dioxide, variable amounts of water vapour and traces of inert gases. The chemical, enzymic and microbial spoilage mechanisms associated with fresh foods may be inhibited by modifying the atmosphere surrounding a food product. (Day, 1992). Proper temperature control to retard the deterioration and ensure the safety of modified atmosphere packaged foods is essential.

The Market For Modified Atmosphere Packaging

4.30.   Over the last few years, the development and commercialisation of modified atmosphere packaging for fresh chilled food products has been most rapid in the UK and France which hold about 40% and 25% of the European market, respectively. "Marketpower" have estimated that 2,000 million modified atmosphere packages were produced in 1990 (Marketpower, 1986 & 1988). By comparison, "Packaging Strategies" have estimated that 3,000 million modified atmosphere packages were produced in Europe in 1988, in contrast to the 500-600 million modified atmosphere packages produced in the US in 1988 (Packaging Strategies 1988).

Modified Atmosphere Packaging Food Applications

4.31.   Established products packaged under modified atmosphere include red meats, seafoods, pasta, offal, cheese, bakery goods, poultry, cooked and cured meats, ready meals, dried foods, herbs, fruit and vegetables. Many new value-added products packaged under modified atmosphere have recently appeared in UK retail chill cabinets.

4.32.   Although retail products have tended to attract most of the publicity in relation to modified atmosphere packaging developments, bulk modified atmosphere bag-in-box technology is currently being used for products such as pork primals, poultry and bacon. In addition, this technology is presently being used to gas flush conventionally packaged retail products, such as over-wrapped trays of red meats, into large bag-in-box master packs.

Modified Atmosphere Packaging Gases

4.33. The gas mixture used in modified atmosphere packaging must be chosen to meet the needs of the specific food product, but for nearly all products this will be some combination of carbon dioxide, oxygen and nitrogen. Maintenance of the correct gas mixture within modified atmosphere packs is essential to ensure product safety, quality, appearance and shelf-life extension.

Carbon Dioxide (CO_2)

4.34. CO_2 has bacteriostatic and fungistatic properties and retards the growth of many moulds and aerobic bacteria. The combined negative effects on various enzymic and biochemical pathways result in an increase in the lag phase and generation time of susceptible spoilage microorganisms. However, CO_2 does not retard the growth of all types of microorganisms, and in some cases actually may enhance growth.

4.35. The absorption of CO_2 is highly dependent on the water and fat content of the product. If products absorb CO_2, the total volume inside the package will be reduced giving a vacuum package look known as "pack collapse". CO_2 absorption can also reduce the water holding capacity of meat and seafood products, resulting in an unsightly drip inside the pack. In addition, some dairy and cold-eating seafood products can be flavour tainted, and fruit and vegetables can suffer physiological damage due to high CO_2 levels.

Oxygen (O_2)

4.36. In modified atmosphere packaging, O_2 levels are normally set as low as possible to inhibit the growth of aerobic spoilage microorganisms and to reduce the rate of oxidative deterioration of foods.

Nitrogen (N_2)

4.37. N_2 is effectively an inert gas and has a low solubility in both water and fat. In modified atmosphere packaging, N_2 is used primarily to displace O_2 in order to retard aerobic spoilage and oxidative deterioration.

Other Gases

4.38. Other gases such as carbon monoxide, ozone, ethylene oxide, nitrous oxide, helium, neon, argon, propylene oxide, ethanol vapour, hydrogen, sulphur dioxide and chlorine have been used experimentally or on a limited commercial basis to extend the shelf-life of a number of food products. For example, carbon monoxide has been shown to be very effective at maintaining the colour of red meats, maintaining the red stripe of salmon and inhibiting plant tissue decay. Also, argon has been shown to inhibit the tissue fermentation of sliced tomatoes, reduce the protein breakdown of seafood and extend the shelf-life of prepared fruit (Schvester 1991, Day, 1991, and Powrie 1992).

Modified Atmosphere Packaging Materials

4.39. The main characteristics to consider when selecting packaging materials with regard to modified atmosphere packaging are gas permeability, water vapour transmission rate, mechanical properties, sealing reliability, transparency, and type of package (Day, 1992).

4.40. In most modified atmosphere packaging applications it is desirable to maintain the atmosphere initially incorporated into the package for as long a period as possible. The correct atmosphere at the start will not serve for long if the packaging material allows it to change too rapidly. Packaging materials used with all forms of modified atmosphere packaged chilled foods should have barrier properties.

4.41. Packaging materials used for modified atmosphere packaging must have sufficient strength to resist puncture, withstand repeated flexing and endure the mechanical stresses encountered during handling and distribution. It is essential that an integral seal is formed in order to maintain the correct atmosphere within a modified atmosphere package to prevent external microbial contamination and air dilution of the contained gas mixture. Therefore, it is important to select the correct heat sealable packaging materials and to control the sealing operation.

Modified Atmosphere Packaging Machinery

4.42. Depending on the product to be packaged and the desired type of pack, five major types of modified atmosphere packaging machinery are commercially used (Day, 1992). Horizontal form-fill-seal (HFFS) and vertical form-fill-seal (VFFS) machines are capable of making flexible pillow packs from only one reel of packaging film. In HFFS and VFFS machines the air from the pack is removed by a gas flushing technique which dilutes the air surrounding the food product before the package is sealed. HFFS machines are mainly suitable for the retail modified atmosphere packaging of bakery products, snack foods and cheese whereas VFFS machines are mainly suitable for the retail modified atmosphere packaging of prepared salads, grated cheese and dried products.

4.43. Thermoform-fill-seal (TFFS) machines produce packages consisting of a thermoformed semi-rigid tray which is hermetically sealed to a flexible lidding film. TFFS machines use a compensated vacuum technique whereby the air is removed by pulling a vacuum on the air within the package and then breaking the vacuum with the desired gas mixture. TFFS machines are primarily suitable for the retail packaging of red meats, seafood products, poultry, cured and cooked meat products, cook-chill foods, pizza and pasta. As an alternative to HFFS and VFFS machines, TFFS machines can also be used for the retail packaging of bakery products, prepared fruit and vegetables, cheese and dried products.

4.44. Vacuum chamber machines use either preformed bags or trays and utilise the compensated vacuum technique to replace air. These machines can be used for small scale production of vacuum or gas flushed retail and catering packs of primal meat cuts, cooked meats, seafood, poultry and multicomponent ready meals.

4.45. Snorkel-type machines also use the compensated vacuum technique to produce bulk modified atmosphere catering bag-in-box packs. Alternatively, they can gas flush conventionally packaged retail products, such as over-wrapped packs of red meat, into large market packs. Snorkel-type machines can be used for the packaging of red meats, poultry, seafood, cooked meats, cheese, bakery products, fruit and vegetables.

4.46. The Group has been made aware of the work of the Modified Atmosphere Packaging Club who are funding the production of Guidelines for the Good Manufacturing and Handling of Modified Atmosphere Packed Food Products. These Guidelines will be published and therefore available to industry in general. The Group considers these Guidelines will provide a valuable source of information to industry and advocates that food businesses involved in modified atmosphere packaging follow these guidelines.

HEATING, CHILL TEMPERATURE CONTROLS, DISTRIBUTION AND STORAGE ISSUE

Heating

4.47. Cooked chilled foods may be divided into the following categories according to their method of processing:-
   1. Ingredients (which may include raw, cooked or partly cooked) are vacuum packaged and cooked in hermetically sealed containers, cooled rapidly, stored and transported in the same containers under chill temperatures.

   2. Raw materials are cooked in open vessels such as steam jacketed kettles then pumped while still hot ('hot fill') into bags or other suitable containers, hermetically sealed, cooled rapidly, stored and transported under chill temperatures.

   3. Joints of meat are cooked in vacuum bags or metal moulds, cooled, removed from the bags or moulds, sliced and re-packaged in vacuum or modified atmosphere packs chilled and distributed under chill temperatures.

4.48. Category 1, which includes cuisine sous vide, may be undertaken on a large manufacturing scale using water baths and chilled water cooling canals or continuous equipment such as microwave tunnels. On a smaller scale, for example for use by caterers, equipment claimed by the manufacturers to be designed for use with sous vide is available. Such equipment currently available in the UK uses static ovens with steam as the heating medium.

4.49. The application of thermal processes on an industrial scale is based on long established procedures involving the identification of the slowest heating points both within the heating equipment and the individual food packs, developing cooking schedules designed to ensure that the time and temperature required to give the margin of safety established by experts are consistently applied to all containers during each production cycle. In catering operations such process control may be less well developed.

Chill Temperature Controls

4.50. Chill temperature controls are required for post cook cooling and for maintenance of the chill chain from completion of processing to the point of consumption.

4.51. Since prepared chilled foods are pasteurised and are not ambient stable, rapid reduction in temperature following the thermal process (or following packaging in the case of hot fill) is required to prevent vegetative outgrowth of surviving spores.

4.52. At present there is no specific statutory requirement for the cooling or chilling of vacuum or modified atmosphere packaged foods in the UK. In England and Wales the Food Hygiene (General) Regulations 1970 (HMSO,1970) and the Food Hygiene (Markets, Stalls and Delivery Vehicles) Regulations 1966, (HMSO, 1966) as amended in 1990 and 1991, require cooling of relevant foods to 8°C without avoidable delay following completion of preparation. Storage and distribution must then take place at not more than 8°C (with a 2°C/2 hour tolerance to allow for events such as defrost cycles and other unavoidable reasons). A maximum of 5°C will apply from 1 April 1993 for cooked products intended to be eaten without further reheating. The EC Directive on Meat Products requires ready meals containing meat (whether or not vacuum packaged) to be reduced to 10°C within 2 hours of completion of cooking. The Department of Health Guidelines on Cook-Chill and Cook-freeze catering systems (Department of Health, 1989) recommend cooling to 3°C within 2 hours of completion of cooking but these guidelines do not apply to cook-chill foods, using processes and packaging designed to provide an expected shelf-life of more than 5 days.

4.53. Cooling of vacuum bags in sous vide systems is usually achieved efficiently in chilled water tumbler chillers. In industrial scale installations blast chillers or cryogenic chillers using liquid nitrogen are normally used. There are relatively inexpensive blast chillers currently available which may be suitable for smaller installations.

4.54. The demands placed upon the refrigeration equipment used for storage and distribution will depend upon the maximum temperature specified. As discussed earlier in the Report, in the absence of other controlling factors and unless a thermal process designed to inactivate spores of psychrotrophic *C. botulinum* is applied, the maximum temperature to prevent growth and toxin production by psychrotrophic *C. botulinum* is 3°C.

4.55. Experience in England and Wales with the introduction of statutory temperature controls throughout the manufacturing, catering, distribution and retail trade has highlighted the difficulties experienced by some sectors of the food industry in meeting 8°C and in preparing for 5°C controls in 1993. Although the large supermarkets have invested heavily in the latest equipment and have indicated that 5°C is achievable for most chilled foods. Information from one major retailer indicates that performance in retail cabinets has moved from a mean of 6.5°C (1988) to 3.7°C (1991). Whilst it is encouraging to see this improvement, the existing equipment of the retail chains as well as that of the independent retailers is not designed to maintain 3°C reliably. A large proportion of retail display cabinets would need to have the higher performance required, and the investment and cost of energy would probably be prohibitively high.

4.56. It is possible, however that a dedicated system of storage and control, for example within an in-house catering system, could maintain 3°C as is recommended for those systems covered by the Department of Health Guidelines on Cook-Chill and Cook-Freeze Catering Systems.

4.57. The statutory temperature requirements do not apply to domestic refrigerators. A MAFF funded study of 252 domestic refrigerators found they operated at a mean temperature of 6.04°C with a range of -0.9°C to 11.4°C (Consumer Handling of Chilled Foods, MAFF 1991), and in the USA 20% of domestic refrigerators have been found to exceed 10°C (Van Garde and Woodburn, 1987). Given the extended shelf-lives assigned to vacuum and modified atmosphere packaged products due regard must be given by manufacturers and retailers of the time their products will spend in domestic refrigerators.

4.58.   Because it is currently not possible to attain 3°C temperature controls throughout the manufacturing, catering, retail and distribution chill chain, the Group reiterates its conclusion that the temperatures currently attainable in the chill chain are not low enough to prevent growth and toxin production by psychrotrophic *C. botulinum* in chilled prepared foods.

4.59.   Because of the concern over temperature abuse with vacuum and modified atmosphere packaged chilled foods, the Group recommends that information should be made available to consumers on the correct handling practices with regard to vacuum and modified atmosphere packaged chilled foods. In addition, it is important that the labelling of these products must clearly indicate the maximum temperature at which the product is to be stored and the importance of abiding by the "use by" date marking **(R 15)**. The Group recommends that vacuum and modified atmosphere packaged chilled foods should always carry a "use by" date and welcomes the fact that it is now illegal to sell foods assigned a "use by" date once this has passed **(R 16)**.

4.60.   A WHO HACCP publication has been produced (Bryan, 1992) aimed at providing guidance on the application of HACCP to the preparation and storage of food in homes, food service establishments, cottage industries and street markets. These issues should be addressd in any guidance issued by Government to raise awareness of the *C. botulinum* hazard in chilled foods.

Distribution - Mail Order

4.61.   The practice of selling pre-packaged smoked fish by mail order was highlighted in the Report of the Committee on the Microbiological Safety of Food as one of particular concern. Mail order foods sent to consumers are exempted from the statutory temperature controls to allow industry time to investigate practical means of compliance with necessary temperature controls. The Group examined the results of a small scale scientific study designed to assess the public safety hazard from mail order foods (The Microbiological Status of Some Mail Order Foods, MAFF Publications, 1991).

4.62.   This study emphasises the need for effective temperature control for products such as smoked fish unless they are rendered ambient stable. Many of these products are vacuum packaged. The Group understands that the mail order food industry has now formed a trade association which is working on a Code of Practice for its members and is currently undertaking research into coolant gels and other packaging aids. Use of delivery or courier services which can guarantee arrival of the product within a certain time, for example within 24 or 48 hours, are also being investigated. Packaging aids can then be used in conjunction with time limited delivery services to enable vacuum packaged mail order foods to be delivered to consumers under controlled conditions.

4.63.   The Group considers that effective temperature control is equally important in the transport of foods by mail order as with conventionally retailed foods. The Group recommends that foods sent by mail order should comply with the statutory temperature requirements in place for foods delivered by other means. In addition, that those involved in sending food by mail order should ensure that controlling factors, in addition to temperature, are present to prevent the growth of pathogenic microorganisms, including psychrotrophic *C. botulinum* (R 17).

Conclusions

4.64.   Changes in manufacturing, catering and food service practice have presented potential microbiological hazards with vacuum and modified atmosphere packaged chilled foods which need to be fully appreciated. It is clear that the risks of *C. botulinum* overcoming a preservation system increase when the organism is present in high numbers, thus it is important to monitor and control all stages of food production. Mild heat treatments which delay spoilage, or the packaging of foods in vacuum or modified atmospheres to control spoilage, may enhance conditions for growth of *C. botulinum*. Contamination of this type may not be accompanied by obvious signs of spoilage, particularly when psychrotrophic strains are present. Whilst only slow growth will occur in these foods when they are kept chilled, if temperature abuse occurs, growth will be more rapid and this could also allow non-psychrotrophic strains to grow (**C 7**).

4.65.   The Group has considered the existing guidance available to industry and has discussed with industry representatives the need for an awareness of the control measures necessary to ensure the microbiological safety of vacuum and modified atmosphere packaged chilled foods.

4.66.   The Group <u>recommends</u> that there should be a comprehensive and authoritative Code of Practice for the manufacture of vacuum and modified atmosphere packaged chilled foods, with particular regard to the risks of botulism.

Such a document should contain detailed guidance on:

Raw material specification

Awareness and use of HACCP

Process establishment and validation (including thermal process)

Packaging requirements

Temperature control through production, distribution and retail

Factory auditing and quality management systems (including awareness and use of **HACCP**)

Thorough understanding of requirements to establish a safe shelf-life

Thorough understanding of the control factors necessary to prevent the growth and toxin production by psychrotrophic *C. botulinum* in chilled foods

Application of challenge testing

Equipment specifications, particularly with regard to heating and refrigeration

Training   (**R 18**)

TABLE: 13. Findings of a questionnaire issued to Local Authority Environmental Health Departments on 30 August 1991, to identify the extent of Sous-Vide and Vacuum Packing

VACUUM PACKING PREMISES						SOUS VIDE			NUMBER OF QUESTIONNAIRES RETURNED
Packaging before cooking	Packaging pre-cooked food (eg sliced meat)		Packaging raw food			Production units	Premises served	Catering outlets using products made elsewhere	
	Factory	Retail	Meat	Fish	Vegetable				
173(46)	173(76)	298(58)	446(98)	30(21)	23(19)	15(14)	44(8)	15(10)	172

The questionnaire was designed to elicit a rapid response to give the group an idea of the numbers and extent of operations taking place. EHD's were asked to indicate numbers of operations in their areas.

(The figures in brackets are the number of authorities reporting such operations).

39

CHAPTER 5

LEGISLATION AND CODES OF PRACTICE

Introduction

5.1.   Vacuum and modified atmosphere packaged chilled foods are currently controlled in the same general way in UK food law as are foods packaged in other ways. Before discussing the possible need for additional controls for these processes, it is necessary to assess the existing legal position. In addition, the information currently available to industry through codes of practice or guidance notes which cover chilled foods needs to be taken into consideration.

5.2.   Given the single market initiative and the ability for chilled foods to be traded throughout the European Community and beyond, it is necessary to consider legislation or other controls in use or being considered in other countries. Details of all legislation/codes known to the Group are contained in Appendix C, (page 54).

Current UK Legislation and Codes of Practice

5.3.   Under the Food Safety Act 1990 (Section 8) it is an offence to sell or supply for human consumption food which does not meet food safety requirements, eg unfit or contaminated food or that which has been rendered injurious to health (Section 7). The maximum penalties on conviction on indictment for any offence under the Act are an unlimited fine or imprisonment for up to two years, or both. For summary offences under sections 7 (rendering food injurious to health), 8 (selling food not complying with food safety requirements) or 14 (selling food not of the nature, substance or quality demanded) the penalties are a fine of up to £20,000 or imprisonment for up to 6 months, or both. Any other offence on summary conviction carries a maximum fine of up to £5,000 or imprisonment for up to 6 months, or both (Food Safety Act, HMSO 1990).

5.4.   Regulations made under the Act can require (inter alia) that any commercial operations (as defined) to do with food are controlled or prohibited for the purposes of ensuring that food complies with food safety requirements or in the interests of public health. The Act has given powers to Environmental Health Officers (EHOs) who can issue prohibition notices relating to the use of any process or treatment involving the risk of injury to health. In addition, they can order immediate closure of a food business if it poses an imminent risk to health.

5.5.   It is now an offence for food businesses to operate from unregistered premises unless they fall within one of the specified exemptions (for example if already licensed eg dairies; vending machines; packaging of eggs). The registration form requires owners of food businesses to indicate if they are carrying out vacuum packaging. Any change in the nature of the food business has to be notified to the local authority. The Working Group understands that this relates to the type of premises and that a food business starting to use vacuum packaging after registration would not need to notify this use (Food Premises (Registration) Regulations 1991, SI 1991/2825,HMSO,1991).

5.6.   Foods which are microbiologically highly perishable and which are in consequence likely after a short period to constitute an immediate danger to health are now required to carry a ''use by'' date together with any instructions for safe storage. This would include the requirement to keep chilled foods refrigerated. (Food Labelling Regulations 1984 (SI 1984/1305) as amended by the Food Labelling (Amendment) Regulations 1990, (SI 1990/2488), HMSO,1990).

5.7.   The Food Hygiene (General) Regulations 1970 (as amended) lay down requirements for the hygienic production of foods. Foods which will support the rapid multiplication of any pathogenic microorganism present if not adequately controlled by temperature are required to be kept at a maximum temperature of 8°C. Some foods, because of their nature or preparation present a higher risk and those foods will be required to be kept at or below 5°C from 1 April 1993. These temperatures apply throughout the manufacturing, distribution and retail chain in England and Wales. (Food Hygiene (Amendment) Regulations 1990 & 1991; SIs 1990/1431 and 1991/1343, HMSO, 1990&91).

5.8.   All materials and articles which are intended to, or are likely to come into contact with food must conform to the relevant legislation. This would include all vacuum and modified atmosphere packaging, and the machinery used in the processing of the food and the filling of these packages. This legislation is the Materials and Articles in contact with Food Regulations 1987 (SI 1987/1523, HMSO 1987). It is likely that there will shortly be more specific rules for plastic food contact materials and articles.

5.9   Permitted packaging gases are specified in the Miscellaneous Additives in Food Regulations 1980 (SI 1980/1834, HMSO 1980). These include Carbon dioxide, Hydrogen, Nitrogen, Nitrous oxide and Oxygen. If proposed EC directives on food additives are agreed then the regulatory position with regard to some of these gases may change.

5.10.   The Chilled Food Association (CFA) Guidelines for Good Hygienic Practice in the Manufacture, Distribution and Retail Sale of Chilled Foods (CFA, 1989) are subject to an independent audit, by the European Food Safety Inspection Service (EFSIS). EFSIS comprises members of the Meat and Livestock Commission (MLC) and Campden Food and Drink Research Association. Membership of the CFA is dependent on achieving the standards laid down in the Guidelines. These Guidelines are currently being updated and will, we understand, include guidance for pasteurisation heat treatments for chilled foods for psychrotrophic *C. botulinum* in addition to those for *L.monocytogenes*.

5.11.   The use of vacuum packaging for meat and meat products is covered in codes of practice issued by the British Meat Manufacturers Association (BMMA). The BMMA also have a system of independent auditing of members, run by the MLC, with membership of the Association being dependant on compliance with BMMA Standards.

5.12.   The risks associated with vacuum packaged fish are recognised in guidance produced both by industry (British Trout Association) and most recently by Government in the leaflet "Microbiological Safety of Smoked Fish" (MAFF Publications, 1991).

Current EC and International Legislation and Codes of Practice

5.13.   The only legislation which we could find which has been designed specifically to encompass sous vide products is that of France. The requirements of the French law are set out in Appendix C, (page 54).

5.14.   Other countries, as in the UK, have general hygiene and food safety legislation but do not, we believe, set detailed statutory rules governing the production of extended-life chilled foods. Several countries and organisations such as CODEX are in the process of considering codes of practice for industry covering extended-life chilled foods. This shows a growing awareness of the potential hazards of improperly managed chilled food products and, in particular the potential hazard presented by psychrotrophic *C. botulinum*. Many of the draft codes contain similar provisions which focus on *L.monocytogenes* and *C. botulinum* as the two microorganisms of most concern in chilled foods. HACCP systems are recommended and many countries are drawing up D and Z values for psychrotrophic *C. botulinum*.

5.15.   The European Community is currently completing work under the single market initiative. There are a number of sector-specific hygiene Directives which are designed to harmonise the rules governing the production and placing on the market of products such as red meat, fish and shellfish, and milk and milk products. Much of this legislation sets general hygiene rules and chill temperature controls for production and transport of products. Product areas such as vacuum packaged meat and prepared meat meals (including sous vide meat meals) are covered by this legislation which will be implemented in the UK. For products not covered by these sector specific measures, and for retail and catering establishments, general hygiene principles will be laid down in the draft EC Hygiene of Foodstuffs Directive currently under discussion in Brussels. The basic provisions of this Directive are broadly similar to those already in UK law and the importance of the HACCP approach is recognised. This Directive, together with harmonised food inspection under the Official Control of Foodstuffs Directive, is designed to provide the basis for ensuring microbiological food safety throughout the Community.

5.16.   Approaches being considered in the USA covering extended-life chilled foods include banning the vacuum packaging of fish and fish products at retail level because of their higher levels of *C. botulinum* type E spores and because of the history of botulism outbreaks associated with fish.

5.17.   The Australian Quarantine and Inspection service (AQIS) is drawing up a code of practice to cover sous vide products and hot fill systems. Detailed specifications are laid down for equipment and buildings.

41

Guidance will be given on time/temperature lethality for psychrotrophic *C. botulinum* and *L.monocytogenes*.

Proposed UK Legislation and Codes of Practice

5.18. The Group considered the following mechanisms which could be used to ensure the safety of vacuum and modified atmosphere packaged foods.

Self-Regulation

5.19. Self-regulation (sometimes known as auto-control) normally involves the adoption by an industry sector of voluntary, i.e. non-statutory codes of practice. These have no legal force. They may however, be taken into account by the courts and by enforcement officers, and will often be quoted in contracts by the purchasers of the products of a particular plant.

5.20. Non-statutory codes of practice can attempt to explain why certain practices should be followed in a particular industry sector, which legislation cannot do. A code of practice can be advisory rather than prescriptive in form, allowing flexibility of operation. Codes can also be updated more easily than legislation to take account of changing industry practices. Compliance with non-statutory codes of practice cannot be enforced, but only recommended.

Registration

5.21. Registration means that proprietors of food businesses in Great Britain are required to apply for their premises to be entered in the register of the enforcement authority (the local authority in whose area the premises are situated). The objective of registration is to ensure that local authorities know about the food premises in their area so that they may inspect them according to the degree of risk they represent. The method of processing being carried out by a business is one factor in determining the frequency of inspection. Inspection is important, since without it there can be no control - bad practices are unlikely to be detected and the availability of advice would be limited.

Licensing

5.22. Ministers have been given powers under the Food Safety Act 1990 which would enable them to require that premises used for the purposes of particular food businesses be licensed before they are allowed to operate. It is possible to stipulate that certain conditions must be fulfilled before a licence can be granted and to prohibit the use of any food premises except in accordance with the conditions of a licence. Where licensing was required it would be an offence to operate in breach of the terms of a licence. A licence could be withdrawn if its conditions were breached and such withdrawal would prevent the business from operating.

5.23. The main difference between licensing and direct imposition of conditions by statutory regulation is that licensing vests discretion in the licensing authority. It is the licensing authority which has the power of prior approval and which may also withdraw a licence for a breach of its conditions. Licensing may be operated either by central or by local government. In cases where there is a considerable number of outlets, it is most likely that local government would be responsible for licensing.

5.24. The terms and conditions of the licence may be stipulated from the centre, or may be decided by the local authority itself. The terms and conditions of the licence need to be clear and enforceable. The less specific they are, the greater is the possibility of divergent interpretations.

5.25. The Group considered the question of whether licensing should be introduced for processes of concern. From the evidence to hand the Group could not single out specific food processes of concern for which licensing should be introduced which could be adequately defined without excluding other, similar processes. The alternative of singling out certain food categories, as identified in Chapter 3, would be impractical and bureaucratic, given the numbers of food businesses currently using vacuum packaging.

Regulation of Specified Processes

5.26. Regulations made under the Food Safety Act 1990 may lay down practices which are required, regulated or prohibited in the use of any process or treatment in the preparation of food for the purposes of ensuring that food complies with food safety requirements or in the interests of public health. (As an example, the Food Hygiene (Amendment) Regulations 1990 and 1991 lay down temperature controls for certain foods). With all legislation, the important factor is that the regulations should be clear and enforceable. For example, if regulations lay down certain values to be met e.g. pH values of particular foodstuffs, these must be measurable both by industry and by enforcement authorities.

5.27.　　Regulations made under the Food Safety Act 1990 would need to define the specific criteria for a process to be permissible. However, it would not be possible to specify every safe combination of technical criteria with regard to *C. botulinum*. If the regulations were to be restricted to certain processes, such as cuisine sous vide, these would need to be defined so that similar processes of concern were not excluded.

Conclusions and Recommendations

5.28.　　Registration is now in place and it is both a right and a requirement of all food businesses to register their premises. The Group believes this should prove a useful aid to enforcement authorities in identifying businesses carrying out vacuum packaging processes. The Group recommends consideration should be given to the requirement to notify a change of nature of business under the Food Premises (Registration) Regulations 1991 being extended to include the move into use of vacuum/modified atmosphere packaging **(R 19)**.

5.29.　　The use of non-statutory codes of practice is well established in the food industry. The Department of Health Code of Practice on Low Acid Canning (HMSO, 1981) has proved successful in guiding that sector. The Group notes that this code is currently being updated and expanded to include processes such as aseptic packaging. The sector being considered by this Group is analogous to that covered by this code and we have recommended that a similar, authoritative code should be available to the industry involved in vacuum and modified atmosphere packaging of chilled foods (para 4.66, page 38).

5.30.　　The need for additional controls on businesses using processes of concern must be considered in the light of the powers already available to enforcement authorities under the Food Safety Act 1990. The Group considers that registration of food premises is a significant step towards identifying businesses using processes of concern and concludes that enforcement authorities have the powers necessary to ensure the production of safe food **(C8)**.

5.31.　　Industry and enforcement officers may lack the knowledge and awareness of those processes and those food categories which present the greatest risk of growth and toxin production by psychrotrophic strains of *C. botulinum* in chilled foods. The production of a definitive code of practice for the chilled food sector concerned with the risk of psychrotrophic *C. botulinum* in vacuum and modified atmosphere packaged foods should meet any such need **(C9)**.

5.32.　　However, it is essential that such a code of practice is widely adopted and used by industry. It is therefore necessary that the code once produced, will be widely publicised and its use encouraged by enforcement authorities and by central Government. To aid and inform industry until the code of practice is available, the Group recommends that steps are taken to publicise the recommendations of this working group on the control factors necessary to prevent the growth and toxin production by psychrotrophic *C. botulinum* in chilled foods **(R 20)**.

CHAPTER 6

CURRENT RESEARCH IN THE UK RELATING TO FACTORS AFFECTING THE SURVIVAL AND GROWTH OF PSYCHROTROPHIC *CLOSTRIDIUM BOTULINUM* IN FOOD

Research funded by MAFF under the Predictive Food Microbiology Programme.

6.1.   The Predictive Microbiology Programme funded by MAFF aims to produce a series of models which will predict the growth, survival and death of the main foodborne pathogenic bacteria, at conditions relevant to foods and their processing and storage.

6.2.   The MAFF programme started in April 1989 and will finish in April 1994. The programme covers foodborne pathogenic bacteria including *C. botulinum*. It is intended that the database and models produced should enable a self-financing service to be available to industry from 1994 onwards. This is currently referred to as "Food Micromodel" and will be launched on 28 October 1992.

6.3.   A mid term review was prepared by the Management Advisory Group in November 1991. The following account of work on *C. botulinum* under this programme is based largely on this review with modifications in the light of reassessment of the programme at the meeting of Senior Representatives of Contracting Organisations (SRCO) on 29 April 1992.

6.4.   Priority has been given to work on psychrotrophic *C. botulinum*. This reflects the concern that has focused on these bacteria in relation to chilled foods, particularly those that are given a pasteurisation heat-treatment and intended for prolonged storage at refrigeration temperature.

6.5.   It was considered that the information in the literature on heat-resistance and the results of extensive challenge tests mean that the control of non-psychrotrophic, proteolytic strains of *C. botulinum* is reasonably well understood, whereas there is less information on psychrotrophic strains, which are of current concern.

6.6.   The status of work under the MAFF programme is as follows:

6.6.1   A study of the effect of temperatures from 4 to 35°C on the rate of growth of psychrotrophic, type B *C. botulinum* has been completed; a linear relationship was obtained between temperature and the square root of the growth rate. (Graham & Lund 1991). Growth was slightly more rapid than that reported in the literature for strains of type E (Ohye & Scott 1957).

6.6.2   A major study has been made of the combined effect of pH, NaCl and incubation temperatures between 5 and 30°C on growth and survival of strains of psychrotrophic type B from a spore inoculum, and a model based on these results is in the process of being validated by comparison with challenge tests already performed in the UK and against published data.

6.6.3   The combined effect of pH, between 5.1 and 7.0, and lactic acid, up to 1.0%, on growth of psychrotrophic type B at 10°C has been determined. At pH values of 5.6 and below, in addition to the effect of pH there was a marked effect of lactic acid, attributable to the undissociated acid. The work has not been extended to other incubation temperatures because under the modelling programme priority has been given to other work on *C. botulinum*.

6.6.4   A study of the combined effect of incubation temperature, pH and sorbic acid on the probability of growth of psychrotrophic type B *C. botulinum* from a vegetative cell inoculum and the development of a model was part funded by MAFF prior to the programme. This work shows the way in which these factors can be used to provide a high level of safety against the growth of *C. botulinum*. For example, the combination of pH 5.3 and 8°C reduced the probability of growth in 60 days by a factor of $>10^5$, compared with a probability of 1 for growth in 2 days at pH 6.8 and 30°C (Lund et al, 1990).

6.6.5   Concern has been expressed about the extent to which models based on studies of growth could

44

be used to predict toxin production in foods. Following studies at MAFF's Torry Research Station relating toxin production to growth the conclusion has been reached that growth studies can indeed be used to indicate the risk of toxin formation.

6.6.6   In relation to the heat resistance of the spores of psychrotrophic *C. botulinum* in foods, studies have been commissioned by MAFF outside the predictive modelling programme (Gaze & Brown 1990a, 1990b) (see below).

6.6.7   Studies of the heat-resistance of spores of psychrotrophic type B, E and F strains of *C. botulinum* have been funded by industry at the Institute of Food Research (IFR) Norwich and are continuing under a grant from the Department of Health (see below).

6.6.8   It is intended that food validation studies for psychrotrophic *C. botulinum* will be initiated when the growth model is available. Many studies of the growth of psychrotrophic strains in foods have been made at Torry Research Station; these results, and those published in the literature will be used to compare with the growth models. Studies of growth of type E and psychrotrophic type B in food have been funded by MAFF linked with the Predictive Modelling Programme (Gaze & Brown 1991).

6.7.   Experiments at Torry Research Station have investigated growth of psychrotrophic *C. botulinum* inoculated into a wide range of fish, and the effect of smoking, irradiation, packaging atmosphere and storage temperature.

6.8.   The proposed priorities for work on *C. botulinum* (psychrotrophic strains) under the MAFF Predictive Microbiology Programme in 1992-94 are listed as follows:

A = very high priority

1. The effect of pH, and NaCl on heat resistance of spores of psychrotrophic strains at temperatures of 70-95°C.

2. The effect of pH, NaCl, lysozyme and incubation temperature on recovery of heat-damaged spores of psychrotrophic strains.

3. The effect of packaging atmosphere on growth. (ie. the effect of carbon dioxide). The indication from the literature is that CO_2 is not expected to be inhibitory.

4. The effect of inorganic and organic preservatives on growth.

5. Validation of models in foods.

B = high priority

1. The effect of acidulants on survival and growth

C = to be done after work under A and B.

1. The effect of irradiation on survival and growth.

The major aims under this programme can be summarised as follows:

(a) To produce model(s) describing and predicting the combined effects of incubation temperature, pH and NaCl on the rate of growth and the survival of psychrotrophic *C. botulinum* types B, E and F.

(b) To produce models describing and predicting the effect of other preservative factors, including acidulants and carbon dioxide.

(c) To determine the effect of pH and NaCl on D and Z values for spores of psychrotrophic strains in the temperature range 70 - 95°C.

(d) To produce models describing and predicting the effects of storage temperature, pH and NaCl on growth from heat-damaged spores of psychrotrophic strains.

(e) To demonstrate the validity of these models for prediction of survival and growth in foods by experimental studies in foods and by comparison with the results of similar studies in foods reported in the literature.

The detailed planning of the work is subject to regular review in the light of progress and developments in industry and other research projects in this area.

Work funded by MAFF outside the Predictive Microbiology Programme

6.9.   4 MAFF projects have been referred to earlier in this report. These cover:

- the determination of the heat resistance of a strain of psychrotrophic *C. botulinum* type B and a strain of type E in cod and carrot homogenate over the temperature range 70-90°C (Gaze & Brown, 1991);

- the determination of the growth potential of *C. botulinum* type E and psychrotrophic type B in sous vide products at low temperatures (3, 5 and 8°C) (Gaze and Brown, 1991);

- the microbiological status of some sous vide products (Blood, 1990); and,

- an investigation of the heating efficiency of equipment used to cook sous vide products (Sheard, 1991)

Work funded by the Department of Health at the Institute of Food Research (IFR), Norwich Laboratory.

6.10.   Title: Factors affecting the heat-resistance of spores of psychrotrophic strains of *C. botulinum* and the effect of preservation conditions on growth from heat-damaged spores.

6.11.   This project started in August 1991 and is intended to continue until August 1993. The scope of this work is wider than predictive modelling in that it includes work relating to physiological factors involved in spore heat-resistance, germination and growth of vegetative bacteria.

6.12.   The project followed from work funded by industrial companies. The initial work and the project proposal were directed towards assessment and control of the risk posed by psychrotrophic strains of *C. botulinum* in mildly heated, refrigerated foods with an extended shelf-life food, such as sous vide foods.

6.13.   Work under the industrial funding had extended previous information on the extent to which lysozyme can markedly increase the recovery of heated spores of psychrotrophic *C. botulinum* types B, E and F. (Peck, Fairbairn & Lund 1991). This raised the question of the incidence of this relatively heat-resistant enzyme in foods and the implications for growth of spores of psychrotrophic *C. botulinum* after heat-treatment of foods.

6.14.   The aims of the work funded by the Department of Health are: to establish the optimum cultural conditions for detection of spores that survive heat treatment, to investigate the effect of cultural conditions and preparation of spore suspensions on heat-resistance and to compare heat-resistance in phosphate buffer and selected foods, and to determine the effect of preservation conditions, including chill temperature, on growth from sub-lethally heat-treated spores.

Work being undertaken at Unilever

6.15.   At Unilever Research the effect of temperature (5-30°C) and salt concentration (NaCl, 0-3%) on the growth and toxin formation of a cocktail of 9 strains of psychrotrophic *C. botulinum* has been investigated in a factorially designed experiment in a model (strictly anaerobic) broth system. The data has been used to construct a predictive model for the time to toxin formation under these conditions. This model has already been made available to the MAFF Predictive Microbiology Programme and will be published in due course. Studies to examine the effect of nitrite in combination with the above factors in a comminuted meat product are also under way. These data will also be modelled.

6.16.   Thermal inactivation studies at Unilever Research and also under contract with the Institute of Food Research, Norwich have concentrated on the effect of salt (NaCl), pH and lysozyme on the recovery of heat damaged spores. This information has been used to construct a predictive model. A number of validation studies in food products are also under way.

Conclusions

6.17.   The Group considers that the research taking place both within and outside the Predictive Microbiology Programme forms a considerable portfolio of research into the survival and growth of psychrotrophic strains of *C. botulinum* in foods. The Group considers that much of the information industry will require to control psychrotrophic *C. botulinum* in foods can be derived from this programme of work (C10).

6.18.   Studies of the heat resistance of spores of psychrotrophic *C. botulinum* and of the effect of pH and a_w on heat resistance are being undertaken under the current research programmes. The effect of storage temperature, pH, and NaCl on the recovery of heat-damaged spores is also being studied and the extension of this work to include the effect of other factors such as acidulants and nitrite should be considered. The combined effect of storage temperature, pH and NaCl on growth from an unheated spore inoculum has been studied and extension of this work to include the effect of other factors including CO_2, acidulants and nitrite should also be considered (C 11).

6.19.   This report shows that considerable work on psychrotrophic *C. botulinum* and its controlling factors is in progress as part of the MAFF Predictive Microbiology Programme. The Group recommends that the need for further work on this organism be kept under review by the Management Advisory Group of the Predictive Microbiology Programme (R 21).

CHAPTER 7

RECOMMENDATIONS AND CONCLUSIONS

RECOMMENDATIONS

C. botulinum - Epidemiological Information

R1   Given the changes taking place in food production technology, the Group <u>recommends</u> that food manufacturers critically assess all new food processing procedures to ensure elimination of the risk of botulism. (1.12)

R2   The Group <u>recommends</u> that home preservation methods such as home canning or bottling of low acid products such as vegetables and meats, and now home vacuum packaging (except for frozen products), should not be encouraged, given the potential risk of botulism. Techniques such as freezing should be advocated for home use. (1.13).

Psychrotrophic *Clostridium botulinum* - microbiology and control in foods

R3   The Group <u>recommends</u> that in order to build in control factors to prevent the growth of psychrotrophic *C. botulinum* in chilled foods where the organism has not been eliminated or is not sufficiently controlled by some other factor, food manufacturers, caterers and retailers should take account of the time taken for this organism to grow and/or produce toxin at the <u>actual</u> temperatures the food is expected to encounter throughout its shelf-life. This must take into account storage, transport, distribution, retail, catering and domestic stages as appropriate. (2.10)

R4   The Group <u>recommends</u> that in addition to chill temperatures of less than 10°C (statutory chill temperature controls require a maximum temperature of 8°C where applicable), prepared chilled foods with an assigned shelf-life of more than 10 days should contain one or more controlling factors at levels to prevent the growth and toxin production by strains of psychrotrophic *C. botulinum*. (2.11)

R5   The Group <u>recommends</u> to food manufacturers, caterers and retailers that, as part of good manufacturing practice, sound technical evidence should be available in order to demonstrate to enforcement authorities that an assigned shelf-life is appropriate to ensure the microbiological safety of food. (2.13)

R6   The Group <u>recommends</u> that because of the potential for lack of control over the cooking or reheating of foods, cooking or reheating should not be relied upon to destroy any botulinum toxin present in foods, but that other controlling factors should be used in foods susceptible to the growth of psychrotrophic *C. botulinum* in order to prevent growth and toxin production by psychrotrophic strains of *C. botulinum*. (2.20)

R7   The Group <u>recommends</u> that, in addition to chill temperatures which should be maintained throughout the chill chain, the following controlling factors should be used singly or in combination to prevent growth and toxin production by psychrotrophic *C. botulinum* in prepared chilled foods with an assigned shelf-life of more than 10 days:-

  - A heat treatment of 90°C for 10 minutes or equivalent lethality,

  - a pH of 5 or less throughout the food and throughout all components of complex foods,

  - a minimum salt level of 3.5% in the aqueous phase throughout the food and throughout all components of complex foods,

  - an a_w of 0.97 or less throughout the food and throughout all components of complex foods,

  - a combination of heat and preservative factors which can be shown consistently to prevent growth and toxin production by psychrotrophic *C. botulinum*. (2.32)

Controlling Factors of Psychrotrophic *Clostridium botulinum* in Chilled Foods

R8   Foods are classified by the Group as high priority for attention because abuse of shelf-life or chill temperature conditions could result in growth and toxin production by psychrotrophic strains of *C. botulinum*. The Group <u>recommends</u> that the risk of the *C. botulinum* hazard in these chilled foods must be addressed by manufacturers, caterers, and retailers. Examples include smoked salmon and trout, cuisine sous vide products and modified atmosphere packaged sandwiches. (3.12)

Packaging and Manufacturing Processes

R9   The Group <u>recommends</u> that food manufacturers and caterers should ensure that they have a thorough understanding of the operational capabilities of their machines; that the appropriate pouch or tray materials are used; and that the machinery used for establishing a vacuum functions to their required specification. (4.7)

R10   The Group <u>recommends</u> that all vacuum packaging machine manufacturers should include a section in their instruction handbook which alerts the user to the food safety hazards and the risks from organisms such as *C. botulinum*. These issues should also be explained during equipment demonstrations.(4.9)

R11   The Group is concerned about the lack of uniformity shown by sous vide cooking equipment and <u>recommends</u> that the cooking of sous vide products should not be undertaken unless operators have available the technical expertise to ensure that pasteurisation equipment and operating procedures are adequately designed and tested to give a uniform and known heat load to all containers during each cycle. (4.14)

R12   The Group <u>recommends</u> that manufacturers and users of heating equipment used for the production of foods cooked under vacuum ensure that their equipment is capable of achieving the performance they require.(4.15)

R13   The Group <u>recommends</u> that the Steering Group on Microbiological Safety of Food consider undertaking surveillance of vacuum packaged and modified atmosphere packaged foods, including sous vide products, focusing on those areas designated a high priority for attention in Chapter 3 of the report. (4.21)

R14   The Group <u>recommends</u> that enforcement authorities use the data which will be available to them under the Food Premises (Registration) Regulations 1991 to identify the types of food premises using vacuum packaging and similar processes and to target food hygiene inspections and other enforcement activities under the Food Safety Act 1990. (4.28)

R15   The Group <u>recommends</u> that information should be made available to consumers on the correct handling practices with regard to vacuum and modified atmosphere packaged chilled foods. In addition, it is important that the labelling of these products must clearly indicate the maximum temperature at which the product is to be stored and the importance of abiding by the "use by" date marking (4.59).

R16   The Group <u>recommends</u> that vacuum and modified atmosphere packaged chilled foods should always carry a "use by" date and welcomes the fact that it is now illegal to sell foods assigned a "use by" date once this has passed. (4.59)

R17   The Group considers that effective temperature control is equally important in the transport of foods by mail order as with conventionally retailed foods. The Group <u>recommends</u> that foods sent by mail order should comply with the statutory temperature requirements in place for foods delivered by other means. In addition, that those involved in sending food by mail order should ensure that controlling factors, in addition to temperature, are present to prevent the growth of pathogenic microorganisms, including psychrotrophic *C. botulinum*. (4.63)

R18   The Group <u>recommends</u> that there should be a comprehensive and authoritative Code of Practice for the manufacture of vacuum and modified atmosphere packaged chilled foods, with particular regard to the risks of botulism.(4.66)

Such a document should contain detailed guidance on:

Raw material specification

Awareness and use of HACCP

Process establishment and validation (including thermal process)

Packaging requirements

Temperature control through production, distribution and retail

Factory auditing and quality management systems (including awareness and use of HACCP)

Thorough understanding of requirements to establish a safe shelf-life

Thorough understanding of the control factors necessary to prevent the growth and toxin production by psychrotrophic *C. botulinum* in chilled foods

Application of challenge testing

Equipment specifications, particularly with regard to heating and refrigeration

Training

Legislation and Codes of Practice

R19   The Group <u>recommends</u> consideration should be given to the requirement to notify a change of nature of business under the Food Premises (Registration) Regulations 1991 being extended to include the move into use of vacuum/modified atmosphere packaging.(5.28)

R20   The Group <u>recommends</u> that steps are taken to publicise the recommendations of this working group on the control factors necessary to prevent the growth and toxin production by psychrotrophic *C. botulinum* in chilled foods.(5.32)

Current Research Relating to Psychrotrophic *Clostridium botulinum*

R21   This report shows that considerable work on psychrotrophic *C. botulinum* and its controlling factors is in progress as part of the MAFF Predictive Microbiology Programme. The Group <u>recommends</u> that the need for further work on this organism be kept under review by the Management Advisory Group of the Predictive Microbiology Programme. (6.19)

In addition to the recommendations made by the Group, there were a number of other conclusions reached by the Group. These are as follows:

CONCLUSIONS

Clostridium botulinum - Epidemiological Information

C1   In those European countries where the incidence of botulism is higher than in the UK, home preservation of foods such as home curing of meat and bottling of vegetables, is frequently implicated. (1.10)

C2   Of those few reported incidents of botulism associated with vacuum packaged foods it would appear that different types of smoked fish are the most commonly implicated foods. (1.11)

Psychrotrophic *Clostridium botulinum* - microbiology and control in foods

C3   It is recognised that certain bulk vacuum packaged foods are stored for a period under strictly controlled conditions at temperatures between -2°C and 0°C prior to final preparation and repackaging for sale. These temperatures are such as to prevent the growth and toxin production by psychrotrophic strains of *C. botulinum*. However, the Group believes that because it is currently not realistic to maintain temperatures of 3°C or less consistently throughout all parts of the chill chain, it is not acceptable to rely on chill temperatures as the sole method of preventing the growth of psychrotrophic strains of *C. botulinum* in chilled foods with an assigned shelf-life of greater than 10 days. (2.11)

Other Pathogens of Concern in Chilled Foods

C4 With the exception of the anaerobic clostridia, most bacteria which cause foodborne disease grow best in the presence of air and therefore in vacuum and modified atmosphere packaged foods growth will usually be slow. However, risk of growth may be increased as a result of contamination of a sterilised or pasteurised product after thermal processing. This possibility is widely recognised and is the subject of standard procedures in industry where the choice of packaging materials and means of assuring pack integrity, combined with hygienic procedures for handling product post-thermal processing, is designed to prevent re-contamination. Risk of growth may also be increased by abuse of the chill chain. The introduction of statutory temperature controls has lead to a greater awareness of the need to monitor and control temperatures throughout the chill chain (2.47).

Controlling Factors of Psychrotrophic *Clostridium botulinum* in Chilled Foods

C5   The Group's arbitrary assignments of priority for attention to categories of chilled foods are designed to help its work and should not be applied outside the scope of this report. However, they may be used in the preparation of the code of practice recommended in this report to help advise manufacturers, retailers, caterers and enforcement officers of those categories of chilled foods which in the Group's opinion are those most likely to present a risk to the growth and toxin production by psychrotrophic strains of *C. botulinum*. (3.13)

Packaging and Manufacturing Processes

C6   The survey of cuisine sous vide products commissioned by MAFF shows little cause for concern about the microbiological status of cuisine sous vide products currently available in the UK. However, the survey was limited and consideration should be given to a further survey on a larger scale. (4.21)

C7   Changes in manufacturing, catering and food service practice have presented potential microbiological hazards with vacuum and modified atmosphere packaged chilled foods which need to be fully appreciated. It is clear that the risks of *C. botulinum* overcoming a preservation system increase when the organism is present in high numbers, thus it is important to monitor and control all stages of food production. Mild heat treatments which delay spoilage, or the packaging of foods in vacuum or modified atmospheres to control spoilage, may enhance conditions for growth of *C. botulinum*. Contamination of this type may not be accompanied by obvious signs of spoilage, particularly when psychrotrophic strains are present. Whilst only slow growth will occur in these foods when they are kept chilled, if temperature abuse occurs, growth will be more rapid and this could also allow non-psychrotrophic strains to grow (4.64)

Legislation and Codes of Practice

C8   The need for additional controls on businesses using processes of concern must be considered in the light of the powers already available to enforcement authorities under the Food Safety Act 1990. The Group considers that registration of food premises is a significant step towards identifying businesses using processes of concern and concludes that enforcement authorities have the powers necessary to ensure the production of safe food (5.30).

C9   Industry and enforcement officers may lack the knowledge and awareness of those processes and those food categories which present the greatest risk of growth and toxin production by psychrotrophic strains of *C. botulinum* in chilled foods. The production of a definitive code of practice for the chilled food sector concerned with the risk of psychrotrophic *C. botulinum* in vacuum and modified atmosphere packaged foods should meet any such need. (5.31).

Current Research Relating to Psychrotrophic *Clostridium botulinum*

C10   The Group considers that the research taking place both within and outside the Predictive Microbiology Programme forms a considerable portfolio of research into the survival and growth of psychrotrophic strains of *C. botulinum* in foods. The Group considers that much of the information industry will require to control psychrotrophic *C. botulinum* in foods can be derived from this programme of work. (6.17)

C11   Studies of the heat resistance of spores of psychrotrophic *C. botulinum* and of the effect of pH and a_w on heat resistance are being undertaken under the current research programmes. The effect of storage temperature, pH, and NaCl on the recovery of heat-damaged spores is also being studied and the extension of this work to include the effect of other factors such as acidulants and nitrite should be considered. The combined effect of storage temperature, pH and NaCl on growth from an unheated spore inoculum has been studied and extension of this work to include the effect of other factors including CO_2, acidulants and nitrite should also be considered. (6.18)

APPENDIX A

CLOSTRIDIUM BOTULINUM - LABORATORY DIAGNOSIS

1.   Diagnosis of foodborne botulism in the first place depends on clinical observations and treatment should be initiated urgently on the basis of a clinical diagnosis, which may subsequently be confirmed bacteriologically (Gilbert and Willis, 1980). The demonstration of *C. botulinum* toxin in the patient's serum and/or stool, and/or in suspected food, confirms the diagnosis. Cultures of suspected food, and of the patient's faeces and vomitus, may give positive results; but the isolation of the organism is, alone, of relatively little significance in foodborne botulism whereas the detection of toxin provides unequivocal evidence to support diagnosis.

2.   Confirmation of toxic cases of botulism require special facilities for mouse protection tests that are infrequently available in routine diagnostic laboratories. The mouse bioassay provides definitive evidence for both the presence of the toxin and the type. It takes up to 4 days to complete, but is very sensitive, detecting <5 mouse LD50 ml^{-1} in foodstuffs and biological samples. Botulinum toxin may also be detected by means of immunological methods. An ELISA (enzyme-linked immunosorbent assay) method which can detect 5-10 mouse LD50 ml^{-1} of purified toxin and is specific for type A has been described (Shone et al, 1985). The same group has developed the ELISA test for type B which was successfully deployed in the hazelnut yoghurt outbreak, but definitive evidence still relies on the mouse bioassay.

3.   The identification of the *C. botulinum* group in the laboratory is difficult because its biochemical and fermentative properties are unremarkable. The occurence of types B and F in Groups I and II complicates the identification process. The situation is further compounded by the occurrence of clostridial species which are biochemically indistinguishable from them except for the production of neurotoxin. *C.sporogenes*, a saprophyte commonly found on raw food is phenotypically indistinguishable from *C. botulinum* Group I. Moreover there are reports in the literature of nontoxigenic variants of groups I and II (Lynt et al, 1982). The situation has been further complicated by isolation of *C.butyricum* strains which produce type E toxin from two independent cases of infant botulism (Aureli et al, 1986, McCroskey et al, 1986). Two strains of *C.barati*, one from a case of infant botulism (Hall et al, 1985) and one from an adult male who had undergone intestinal surgery (McCroskey et al, 1991) which produced neurotoxin type F have been isolated. The authenticity of the species was confirmed with DNA hybridisation studies (Suen et al, 1988). Recent advances in DNA technology may offer some improvement. Probes designed to detect the neurotoxin gene which could distinguish proteolytic and non-proteolytic type B from the other *C. botulinum* types A - G and from *C.sporogenes* and *C.butyricum* have been described (Szabo et al, 1991).

4. Reference facilities for cases of foodborne botulism and infant botulism in the UK are provided by the PHLS Food Hygiene Laboratory at Colindale.

APPENDIX B

EQUIVALENT 6D TIME/TEMPERATURE COMBINATIONS FOR SPORES OF PSYCHROTROPHIC *CLOSTRIDIUM BOTULINUM*

These data have been calculated using a z value of 9 Centigrade degrees, a reference temperature of 80°C and the lethal rates equation (Stumbo, 1973).

$$L = \log 10^{-1} \frac{T - Tx}{z}$$

where

L = lethal rate

T = the temperature under consideration

Tx = the reference temperature

z = the chosen z value

TEMPERATURE (°C)	LETHAL RATE	TIME (MINS)
70	0.077	1675
71	0.100	1290
72	0.129	1000
73	0.167	773
74	0.215	600
75	0.278	464
76	0.359	359
77	0.464	278
78	0.599	215
79	0.774	167
80	1.000	129
81	1.292	100
82	1.668	77
83	2.154	60
84	2.783	46
85	3.594	36
86	4.642	28
87	5.995	22
88	7.743	17
89	10.000	13
90	12.915	10

APPENDIX C

SUMMARY OF UK, EC AND INTERNATIONAL CODES OF PRACTICE AND EC AND INTERNATIONAL REGULATORY PROVISIONS

1. UK CODES OF PRACTICE/GUIDANCE DOCUMENTS

1.1 British Meat Manufacturers' Association (BMMA) - Standards: 1. for the Production of Bacon and Bacon Joints, 2. Accredited Standards of Good Manufacturing Practice (1991)

Scope - Covers fresh and frozen meat and cured meats such as bacon.

Content - Detailed guidance on hygienic practices throughout handling and preparation of products. Information on clostridia refers to concerns over growth in anaerobic conditions. Standard on bacon covers vacuum packaging and stresses the need for good hygienic practices when carrying out this process. Use of heat shrinkable films and secondary sealing covered as well as modified atmosphere packaging. Storage at temperatures of - 4°C to 5°C recommended.

1.2. British Trout Association - Code of Practice (currently being updated)

Scope - Production of table trout to be sold direct from the farm to processors, wholesale markets, retail fishmongers and caterers.

Content - Covers smoking, processing, packaging, storage and distribution of trout. Detailed requirements for brining given but no strength for brine or salt levels given. Stress is on the need to ensure adequate brining as essential to ensure reduced risk of contamination by food poisoning bacteria such as *C. botulinum.* (revised guidelines to recommend 3.5% salt in the aqueous phase plus temperature controls). Cleanliness to avoid cross contamination stressed, product to be cooled to below 4°C before packaging, to be labelled "store below 4°C", and be kept below 4°C throughout the distribution and sales chain. Recommended shelf-lives are given based on quality; that for vacuum packaged hot-smoked and cold-smoked trout is 21 days.

1.3. Chilled Food Association - Guidelines for Good Hygienic Practice in the Manufacture of Chilled Foods 1989 (under revision)

Scope - Chilled foods ready-to-eat or reheated before consumption. Do not apply to catering operations.

Content - Separates foods into 4 categories and sets specific hygienic requirements for each category. Categories are :

1   Those prepared from raw components.

2   Those prepared from cooked and raw components processed to extend the safe shelf-life.

3   Those prepared from only cooked components.

4   Those cooked in their own packaging prior to distribution.

Specific hygienic requirements, for example for category 4 products, rely on good manufacturing practice, thermal processing sufficient to achieve the target temperature appropriate to the organisms to be controlled, cooling and drying of packs where appropriate and control of time and temperature of distribution and sale. The only specific reference to *C. botulinum* relates to products with an extended shelf-life where the guidelines state that "it must also be demonstrated that the heating process is sufficient to destroy the spores of psychrotrophic strains of *C. botulinum* unless other preservative systems are present." Recommends HACCP. Members subject to independent audit to ensure compliance with the Guidelines.

1.4 Department of Health - Guidelines on Cook-Chill and Cook-Freeze Catering Systems (1989)

Scope - Apply only to catering operations using pre-cooked chilled foods and pre-cooked frozen foods. They do not cover chilled foods produced under conditions of processing and packaging designed to provide an expected shelf-life of more than five days.

Content - Detailed guidance on the hygienic preparation of cook-chilled foods reliant on hygienic preparation, and rapid chilling after cooking to between 0°C and +3°C within 90mins. Cooking should be sufficient to ensure the destruction of non-sporing pathogens. Packing does not cover vacuum or hermetically sealed packs. Advice on management of quality assurance is given with recommendation of use of HACCP.

1.5 Department of Health - Recommended Practices for the Processing, Handling, and Cooking of Fresh, Hot-Smoked, and Frozen Trout (1978)

Scope - Fresh, hot-smoked and frozen trout - handling practices to reduce the risk of contamination with *C. botulinum*.

Content - Details given for the brining and smoking of hot-smoked trout. Product to be chilled to below 4°C after smoking, to be labelled "store below 4°C" and product should not be vacuum packaged. A shelf-life of not more than 5 days is recommended.

1.6 Institute of Food Science and Technology (IFST) - Guidelines for the Handling of Chilled Foods 2nd Edition (1990)

Scope - Perishable foods which, to extend the time during which they remain wholesome, are kept within specified ranges of temperature above -1°C and below +8°C. Covers preparation, storage, and handling of chilled foods to ensure safety and maintain quality.

Content - Detailed coverage of good hygienic practices throughout manufacture (preparation, processing and packaging), distribution and storage of chilled foods. Recommends use of HACCP procedures. Sections cover retail and catering practices. Gives recommended storage temperatures for chilled foods. Notes that vacuum and modified atmosphere packaged foods do not prevent the growth of food poisoning organisms and states the importance of following the manufacturers storage instructions.

1.7 Ministry of Agriculture, Fisheries and Food (MAFF) - Microbiological Safety of Smoked Fish (1991)

Scope - Current fish smoking practices and safety factors which need to be given attention by fish smokers and those involved in the distribution chain.

Content - Covers salting, brining, recommended salt concentrations (minimum 3.5% w/w in the aqueous phase for products to be stored below 10°C), smoking and packaging. Discusses microbiological food safety implications of current practices, stresses the need for rigid temperature control to inhibit growth of pathogenic microorganisms, and gives 10 steps to safer smoked fish production.

1.8 Sous Vide Advisory Committee (SVAC) - Code of Practice for Sous Vide Catering Systems (1991)

Scope - Covers sous vide (La Methode Sous Vide, Cuisine Sous Vide, Cuisine en Papillote Sous Vide) with a shelf-life of not greater than 8 days.

Content - Stresses the need for the highest hygiene standards to be maintained at every step of the operation. Ingredients must be of good microbiological quality and stored under refrigerated conditions. Food prepared and awaiting vacuumisation must be stored at less than 10°C. Food heated as part of preparation must be chilled to 10°C before vacuumisation. Specifications are given for the plastic laminated pouches used and for the types of heating equipment to be used. The heat treatment should be such as to achieve the destruction of vegetative stages of any pathogenic microorganism, and significantly reduce the number of spores of *C. botulinum* (types E and B) present.

The minimum heat treatments recommended to achieve effective pasteurisation of the product assuming post cooking storage at 0 to 3°C are expressed as food core temperatures:

Temp (°C)	Time
80	26 min
85	11 min
90	4.5 min
95	2 min

The need for even heating throughout the product and batch is emphasised as is the need for temperature monitoring. After cooking the food should be chilled to below 3°C within 90mins. Food should be stored between 0-3°C in a dedicated storage area for no more than 8 days including the days of production and consumption. This is regarded as a maximum guideline for non-industrially produced product. Instructions for reheating are given.

Control of the process with particular regard to the critical control points is advocated as the best way to ensure product safety. A regular programme of microbiological testing of end products is recommended but no levels are suggested.

2 EUROPEAN LEGISLATION AND CODES OF PRACTICE

2.1 Benelux - Draft Code for the Production, Distribution and Sale of Chilled, Long-life Pasteurised Meals (1990)

Scope - Production and distribution of chilled pasteurised meals in consumer packaging or bulk packaging with an extended storage life at 0-4°C obtained by minimum heat treatment and hermetic packaging of the meal. Meals are to be eaten after reheating although the Code does not cover details of reheating of catering packs. Storage life at 0-4°C is 3-6 weeks. Code excludes fresh chilled meals with a storage life of less than 10 days and frozen meals.

Content - Specification of hygienic design of premises and equipment. High standard of raw materials, packaging materials etc. required. Assembly of ready meals should take place either at ≥70°C or ≤10°C and in a "high care room". Pasteurisation should take place as rapidly as possible. Pasteurisation times/ temperatures should be calculated according to the target organism. eg 2 mins/70°C *L.monocytogenes*; or to inactivate spores of psycrotrophic non-proteolytic *C. botulinum* types B,E and F, 10 mins/90°C produces a reduction of 6D. Lethal values for *Listeria monocytogenes* and *C. botulinum* are to be included in the annex to the Code. Meals pasteurised by use of the lethal value for *C. botulinum* are to be called "highly pasteurised". Separate requirements are given for meals cooked in hermetically sealed containers and those cooked and then sealed. A second pasteurisation to the *Listeria monocytogenes* lethal value is given to the latter.

2.2 Council Directive 89/397/EEC of 14.6.1989 on the Official Control of Foodstuffs (European Commission 1989)

Scope - Covers all stages of the production and distribution chain.

Content - Lays down general rules for enforcement authorities when inspecting foodstuffs. The Directive requires that inspections shall be carried out regularly and cover all stages of the production, manufacturing and distributive chain. Control includes microbiological sampling and analysis.

2.3 Council Directive 92/5/EEC of 10.2.92 on health conditions of the production and marketing of meat products and certain other products of animal origin ,amending and updating Directive 77/99/EEC. (European Commission 1992)

Scope - Production, manufacture and distribution of meat products and certain other products of animal origin.

Content - Products covered include meat products prepared from red, white, game and rabbit meat, and other products of animal origin (including meat extracts, gelatine, intestines and melted animal fat). Packaging of non-ambient stable products to have indication of appropriate transportation and storage temperatures and use by date for microbiologically perishable products. Production establishments shall be checked to ensure the microbiological and hygiene condition of the products of animal origin. Temperature controls are laid down for meat based prepared meals, rendered animal fats, greaves and by products and for other products.

2.4 Council Directive 91/497/EEC of 22.7.91 laying down health conditions for the production and the placing on the market of fresh meat (European Commission 1991)

Scope - Introduces a system of common public health rules for all fresh meat produced in the EC after 1992. Covers production and distribution.

Content - Meat to be chilled and transported at +7°C and offal 3°C. Plant operators must carry out microbiological checks on tools, equipment and, if necessary meat.

2.5 Council Directive 92/46/EEC of 16.6.1992 laying down health rules for the production and laying on the market of raw milk, heat treated milk, and milk based products. (European Commission 1992)

Scope - Covers raw milk, heat treated milk and milk based products throughout the production and distribution chain.

Content - Sets out microbiological standards to be achieved in the production of raw milk, heat treated milk and milk based products for human consumption, from cows, ewes, goats and buffaloes. Also sets rules with reference to other EC legislation where appropriate for standards of hygiene to be met by and maintained in production holdings, processing plants and manufacturing establishments; packaging, transport and storage; and labelling. Sets requirements relating to the maximum permitted temperature for milk and milk products, through the collection, production, storage and distribution chain.

2.6 Council Directive 91/493/ EEC of 22 July 1991 laying down health conditions for the production and the placing on the market of fishery products (European Commission 1991)

Scope - Covers raw fish and fish products throughout the production and distribution chain.

Content - Harmonises controls for handling and treatment of all fish and shellfish up to, but excluding retail stage. Specifies hygiene conditions for premises, equipment, staff, processing and handling of fish. Temperature controls during storage and transport require that fresh/thawed fishery products and cooked/chilled crustacean and molluscan shellfish must be kept at the temperature of melting ice. Processed products must be kept at temperatures specified by manufacturers.

2.7 France - Ministry of Agriculture - Hygiene Regulations Concerning the Preparation Distribution and Sale of Pre-Cooked Ready Meals (1974)

Scope - Ready to eat meals, cooked or pre-cooked with a shelf-life at least 24 hours after cooking and up to 5 days, ie 6 days in total. Frozen and chilled meals. Excludes sterilised products, cooked pork meats and salt food products. Production, storage, transport and distribution covered.

Content - Registration of premises preparing ready to eat meals. Hygienic specification of buildings and equipment, specification of refrigeration facilities to store frozen meals at -18°C (max) and chilled foods between 0°C and +3°C. Cooling time of less than 2 hours after cooking to reach a core temperature of 10°C (max). Storage must then be at less than or equal to 3°C. Storage temperature must be on the pack plus a health mark, the date of cooking and "best before". Reheating requirements are given, and thawing before sale prohibited. Microbiological limits for end product and methods of analysis set (Salmonella, Staphylococcus, Coliforms, *E.coli*, Sulphate reducing anaerobes, Mesophilic aerobic bacteria).

2.8 France - Ministry of Agriculture - Extension of the Shelf-life of Ready to Eat Meals (1988)

Scope - Chilled meals as in 2.1 above.

Content - Operators of registered premises can apply for authorisation to extend the shelf-life of particular chilled ready to eat meals whose heat treatment is carried out in the final packaging for up to 42 days. Products have to remain within the microbiological criteria set out in the 1974 Regulations for up to 48 hours after the expiry of the shelf-life with product kept at 3°C. In addition manufacturers have to carry out checks to ensure compliance with a safety margin as follows:

21 day shelf-life - 14 days at +4°C, then 7 days at +8°C

42 day shelf-life - 28 days at +3°C, then 14 days at +8°C

Core temperatures for cooking are set of 57°C-65°C (assessed by an accredited laboratory); Core temperature of 65°C using a pasteurisation value of at least $VP_{70}10=100$, shelf-life of up to 21 days; Core temperature of 70°C using a pasteurisation value of at least $VP_{70}10=1000$, expiry date of no longer than 42 days.

Examples of reliable measures of core temperatures are given as follows:

Temperatures	Time	VP$_{70}$10
69°C ± 2°C	50 min	25 to 63
71°C ± 2°C	50 min	40 to 100
74°C ± 2°C	40 min	63 to 159
75.3°C ± 2°C	295 min	631 to 1,584

The recommended minimal heat treatment for "sous vide" products is 70°C/40 min (VP = valeur pasteurisatrice - pasteurisation value = 40 or P70 = 40), which will achieve a 13 log cycle reduction of *Streptococcus faecalis*.

2.9 France - Syndicat Nationale des Fabricants de Plats Prepares (SYNAFAP)- Recommendations for Good Manufacturing Practices for Prepared Refrigerated Meals (1989).

Scope - as 2.8 & 2.9

Content - Detailed code of practice for preparation of prepared refrigerated meals. Covers general hygiene requirements for buildings, personnel. Pasteurisation values and cooking values are given based on reduction of *Streptococcus faecalis*.

Both pasteurisation and cooking should be completed by rapid chilling to ≤3°C followed by chilled distribution (0-4°C) during which the product temperature should not exceed 6°C and sale at ≤4°C. The Code recognises that existing retail cabinets perform between 4 and 7°C. Whilst accepting 7°C for the present, a target of 3 years is given to bring them up to standard. Labelling requirements are given and to prevent chilled meals being stored by consumers for too long in refrigerators which cannot meet this requirement "consume within one week of purchase; observe the latest consumption date".

Storage life challenge testing requirements follow the French guidelines re storage times/temperatures. Microbiological specifications are given to be applied to the meal after production and also for the meal on the latest consumption date.

3 INTERNATIONAL

AUSTRALIA

3.1 AQIS - Draft Code of Hygienic Practice for the Manufacture of Sous Vide Products (Canberra 1991)

Scope - Covers cuisine sous vide products destined for chill or for frozen storage. Also covers "hot fill".

Content - Sets out minimum hygiene standards for buildings, personnel, processing and distribution. Guidance on microbiological status of ingredients including Total Plate Counts; packaging materials. Approval of process/product required. Products have to undergo a 12D process for a chosen target microorganism - guidelines for *L.monocytogenes* and *C. botulinum* thermal inactivation studies are given. Equipment used for thermal processing must be operated in accordance with and meet the legal standards which apply to low acid canned foods. Requires cooling to a core temperature of 7°C within 4 hours. Details labelling instructions, retail storage temperatures of ≤3°C or ≤8°C depending on thermal treatment with the onus on the manufacturer to ensure that these can be met. Details application of HACCP to sous vide production. Calculation of D and Z values are given for Listeria but in this draft, not for *C. botulinum*. Code will probably become a Standard and have the force of law.

CANADA

<u>3.4 Agri-Food Safety Division Agriculture Canada - Code of Recommended Manufacturing Practices for Pasteurised/Modified Atmosphere Packed/Refrigerated Food (1990)</u>

<u>Scope</u> - Covers meat and meat products which are pasteurised, packaged either prior to or following pasteurisation, under modified atmospheres in hermetically sealed containers, and which require refrigeration (-1 to 4°C) throughout their shelf-life. Pasteurisation is intended to achieve a specified level of decimal reductions in the number of target microorganisms. Designed to reduce and/or eliminate vegetative forms but not sufficient to affect spores of spore-forming organisms and therefore does not lead to shelf stable products.

<u>Content</u> - Based on HACCP as a preventative approach leading to a higher degree of confidence in product safety than traditional safety assurance programs and designed to assist and encourage compliance with the relevant food safety legislation. Model of HACCP approach given, also scenarios of main processes used to produce the products in question. Factors affecting microbiological contamination identified. For each step of the process risk analysis should be carried out.

Recommended manufacturing practices are given for raw materials, packaging, hygienic practices i.e. plant layout equipment, cleaning, personnel. Preparation and packaging is covered with product to be kept below 4°C at all times. Pasteurisation is defined and the total pasteurising value is calculated as:

1. The Decimal Reduction Time ($D_{T°}$) defined as the time required at a reference temperature $T°$, to reduce the microbial population of the target organism by 90%, i.e. 1 log cycle.

2. The number of Decimal Reductions (n) defined as the number of log cycles or decimal reductions of the target organism to be obtained post heat treatment.

3. The Temperature Interval Z or Z value defined as the increase (or decrease) in temperature necessary to reduce (or increase) the decimal reduction time by a factor of 10 or 1 log cycle.

The time necessary to obtain a given number of decimal reductions (n) at the reference temperature ($T°$) is called the Minimum Pasteurising Value ($Pm^ZT°$) and is given by the equation $Pm^ZT° = n\,D_{T°}$. Examples are given of how to work out the decimal reduction time for different organisms. The French regulations are quoted as examples.

Microbiological criteria are set for the product after processing, but not at the end of shelf-life. The criteria given include: absence of Salmonella spp, *E.coli*, *Staphylococcus aureus*, and *L.monocytogenes*; a coliform count of less than 10^2 cfu per gram and an APC of less than 10^2 cfu per gram.

CODEX ALIMENTARIUS COMMISSION

<u>3.5 CODEX Alimentarius Commission - Recommended International Code of Practice on General Principles of Food Hygiene (1985)</u>

<u>Scope</u> - General food hygiene matters through food preparation and processing.

Other CODEX work in hand of relevance to this report include:

General principles of the application of HACCP to CODEX Codes of Hygienic Practice;

A code of hygienic practice for cooked and pre-cooked foods in mass catering;

A code of hygienic practice for aseptic food processing and packaging systems;

Consideration of the occurrence of *L.monocytogenes* in foods.

Consideration of a French document on refrigerated packaged foods with extended shelf-life.

<u>Content</u> - CODEX Codes of practice and CODEX Standards are advisory in nature. They set out good manufacturing practice to be followed in specific sectors. The Codes and Standards are regularly reviewed and updated.

UNITED STATES OF AMERICA

3.6 FDA/Maryland State Health Department - Recommendations for Vacuum Packaging of food in Retail Stores

Scope - Recommends that retailers do not carry out vacuum or modified atmosphere packing of foods. Licences can be granted permitting vacuum packaging or modified atmosphere packaging of foods at retail of foods which will not support the growth of *C. botulinum*.

Content - Restricts foods to be vacuum packaged to those with a water activity below 0.93; pH 4.6 or less; cured products; products with a high level of competing bacteria such as raw meat, raw poultry. Fish and fish products (except frozen) may not be vacuum packaged at retail. Sets maximum 10 day shelf-life for all refrigerated vacuum packaged products which must be kept at or below 45°F. Operators must certify to the regulatory authority that the individual responsible for the vacuum packaging operation understands the equipment, the procedures and the concepts required for safe vacuum packaging.

3.7 Illinois State Department of Health - Retail Guidelines for Refrigerated Foods in Reduced Oxygen Packages (Final Draft).

Scope - Establishments which process, prepare, store, handle or sell or offer for sale at retail chilled foods in reduced oxygen packaging are required to apply to the appropriate regulatory authority for approval. Such approval granted on a product by product basis. Covers hot fill, vacuum packaging, sous vide, modified atmosphere packaging and controlled atmosphere packaging.

Content - Producers need to describe processing, packaging and storage procedures. They must identify the critical control points in the procedure with a description of how these will be monitored and controlled. Hygienic requirements for production premises and personnel are laid down.

Products must, because of their characteristics, present a barrier to the growth of *C. botulinum*. Refrigeration at less than +45°F (7.2°C) is the recognised primary barrier but products must also possess one or more characteristics such as a_w below 0.91; pH less than 4.6; high levels (unspecified) of non-pathogenic competing organisms; meat or poultry with nitrite levels of at least 120 ppm and a minimum brine content of 3.5%; product may alternatively be frozen. Raw or processed fishery products in reduced oxygen packs may not be processed for sale at retail level unless held frozen.

Labelling requirements are stipulated together with "Use By" dates. These latter must not exceed 14 days from retail processing.

3.8 U.S. National Advisory Committee on Microbiological Criteria For Foods - Recommendations for Refrigerated Foods Containing Cooked, Uncured Meat or Poultry Products that are Packaged for Extended Shelf-life and that are Ready to Eat or Prepared with Little or No Additional Heat Treatment (1990)

Scope - Refrigerated foods containing cooked, uncured meat or poultry products that are packaged for extended shelf-life and that are ready to eat or prepared with little or no additional heat treatment. Separate Group to report on seafood and possibly pasta and vegetable products. Excludes products already under regulatory control, for which there is extensive commercial experience and a favourable record of safety (eg cooked beef, roast beef products).

Content - Require producers to operate under a verified HACCP program. This must address the control of *C. botulinum*, *L. monocytogenes* and Salmonella spp. Processes should have to be such as to achieve a minimum 4 log reduction for *L. monocytogenes* and to ensure that toxin production by non-proteolytic and proteolytic *C. botulinum* is controlled. Licensing of premises proposed in non-federally inspected premises.

Types of processing broken down into 9 groups with examples of typical foods given eg: sous vide: - raw ingredients - pre-cook (optional) - formulate - vacuum package - pasteurise - chill* - distribute. * denotes chilling from that point until sale. These nine groups further reduced by grouping into 3 categories:-

    1. assembled and cooked - heat treatment applied after packaging should kill a portion of the normal spoilage bacteria but not *C. botulinum*. Time/temperature control procedures must ensure ultimate safety.

    2. cooked and assembled, and

3. assembled with cooked and raw ingredients - cooked ingredients must receive a heat treatment to destroy non spore-forming bacteria. Essential to ensure cooked products are not contaminated by pathogens during assembly and packaging. Raw ingredients and the assembly process must be controlled to prevent recontamination.

The use of containers traditionally used for shelf stable foods not recommended for "keep refrigerated" products. Packaging systems must be monitored and controlled. Product distribution must be addressed in the HACCP plan. Labelling and shelf-life formulation are addressed.

Appendices give cooking and cooling parameters; information on packaging systems; examples of the application of HACCP principles; D values for *L.monocytogenes*; etc.

3.9 National Food Processors Association - Microbiological and Food Guidelines for the Development, Production, Distribution and Handling of Refrigerated Foods (1989)

<u>Scope</u> - Refrigerated perishable processed foods.

<u>Content</u> - Covers product and process development; HACCP etc. Recommends one or more safety factors in addition to refrigeration be incorporated in refrigerated food formulae. In addition challenge testing designed and supervised by experts in the field should be carried out. Packaging should be tamper proof and should be such as to avoid confusion with shelf stable products. Foods grouped into two for labelling purposes : A - highly perishable, packaged, processed foods which must be refrigerated for safety reasons; B - products intended to be refrigerated which do not pose a safety hazard if temperature abused. Group A should be labelled "Important - must be kept refrigerated" and "must be used by". Temperature of 4.4°C to be maintained until end of shelf-life or purchase by consumer.

3.10 New York State - Requirements for Vacuum or Modified Atmosphere Packaging at Retail (1989)

<u>Scope</u> - Recommends that retailers do not carry out vacuum or modified atmosphere packing of foods. Licences can be granted permitting vacuum packaging or modified atmosphere packaging of foods at retail of foods which will not support the growth of *C. botulinum*.

<u>Content</u> - Restricts foods to be vacuum packaged or modified atmosphere packaged to those with a water activity below 0.93; pH 4.9 or less; cured products; products with a high level of competing bacteria such as raw meat, raw poultry. Fish and fish products (except frozen) may not be vacuum packaged at retail. Smoked fish or poultry may not be vacuum packaged at retail. Sets maximum 14 day shelf-life for all refrigerated vacuum packaged products at a temperature of 45°F. Any vacuum packaging at retail must be carried out under licence.

BIBLIOGRAPHY

SECTION 1: GOVERNMENT PUBLICATIONS

HMSO, - The Food Hygiene (Markets, Stalls and Delivery Vehicles) Regulations 1966, (Statutory Instrument 1966 No 791), HMSO 1966.

HMSO, - The Food Hygiene (General) Regulations 1970, (Statutory Instrument 1970 No 1172), HMSO (reprinted 1989)

HMSO, - The Miscellaneous Additives in Food Regulations 1980, (Statutory Instrument 1980 no 1834), HMSO 1980.

HMSO, - Food Hygiene Codes of Practice - 10: The Canning of low acid foods (HMSO London 1981)

HMSO, - The Food Labelling Regulations 1984, (Statutory Instrument 1984 no 1305), HMSO 1984.

HMSO, - The Materials and Articles in Contact with Foods Regulations 1987 - (Statutory Instrument 1987 No 1523), HMSO 1987.

HMSO, - Home Preservation of Fruit and Vegetables - Agriculture and Food Research Council, 14th edition, 1989.

HMSO, - The Preservatives in Food Regulations 1989 - (Statutory Instrument 1989 No 533), HMSO 1989.

HMSO, - The Food Safety Bill Second Reading Debate, House of Lords Official Report, 5.12.1989, Col 813, para. 3 (of 5).

HMSO, - The Food Safety Act 1990, Chapter 16, HMSO London 1990.

HMSO, - The Food Hygiene (Amendment) Regulations 1990 - (Statutory Instrument 1990 No. 1431), HMSO 1990.

HMSO, - The Food Labelling (Amendment) Regulations 1990, (Statutory Instrument 1990 No. 2488), HMSO 1990.

HMSO, - The Food Hygiene (Amendment) Regulations 1991 - (Statutory Instrument 1991 No. 1343), HMSO 1991.

HMSO, - The Microbiological Safety of Food - Part I (Report of the Committee on the Microbiological Safety of Food, Chairman: Sir Mark Richmond), London: HMSO, 1990

HMSO, - The Microbiological Safety of Food - Part II (Report of the Committee on the Microbiological Safety of Food, Chairman: Sir Mark Richmond), London: HMSO, 1991.

HMSO, - Food Safety Act 1990 - Code of Practice No. 9: Food Hygiene Inspections, London: HMSO 1991.

HMSO, The Food Premises (Registration) Regulations 1991 - (Statutory Instrument 1991 No. 2825), HMSO 1991.

Ministry of Agriculture, Fisheries and Food, (News Release 416/89), - Government Proposes to Regulate Vacuum Packed Foods, 26.10.1989.

Ministry of Agriculture, Fisheries and Food, Consumer Handling of Chilled Foods: A Survey of time and temperature conditions, MAFF Publications, 22.10.91, PB 0682.

Ministry of Agriculture, Fisheries and Food, Time - Temperature Indicators: Research into Consumer Attitudes and Behaviour, MAFF Publications, 22.10.91, PB 0685.

Ministry of Agriculture, Fisheries and Food, The Microbiological Status of Some Mail Order Foods, MAFF Publications, 7.11.91, PB 0707.

Ministry of Agriculture, Fisheries and Food, Microbiological Safety of Smoked Fish, MAFF Publications 7.11.91, PB 0708.

Ministry of Agriculture, Fisheries and Food, (Food Safety Directorate News Release FSD 5/92), - Government Accepts Advice on OHMIC Heating and Geranium, 28.1.1992.

Dept. of Health - A small outbreak of suspected botulism. Annual Report of the Chief Medical Officer, 1948, pp 90-91.

Dept. of Health - Recommended Practices for the Processing, Handling and Cooking of Fresh, Hot-Smoked, and Frozen Trout, 1978.

Dept. of Health - Chilled and Frozen - Guidelines on Cook-Chill and Cook-Freeze Catering Systems, 1989.

Dept. of Health, MAFF, COI, - (Leaflet) HACCP: Practical Food Safety for Businesses, HMSO, October 1991.

SECTION 2: PAPERS AND DOCUMENTS

Adams C.E. (1991) - Applying HACCP to Sous Vide Products, Food Technology, 148-151

Agriculture Canada, - Canadian Code of Recommended Practices for Pasteurised/Modified Atmosphere Packaged/Refrigerated Food, Agri-Food Safety Division, Agriculture Canada, March 1990.

Allen R.L. (1991) - Sous vide's popularity still lukewarm in U.S., Nation's Restaurant News, 16.12.91, Vol 25, No 49.

Arnon, S.S. (1980), -.Infant botulism. Annu. Rev. Med. 31, 541-560.

Aureli, P., Fenicia,L., Pasolini, B., Goinfrancheschi, M., McCroskey, L.M., and Hatheway C.L. (1986) - Two cases of type E infant botulinsm in Italy caused by neurotoxingenic Clostidium botulinum. J.Infect. Dis.154, 207-211.

Australian Quarantine and Inspection Service (AQIS), Code of Hygienic Practice for the Manufacture of Sous Vide Products, AQIS, Department of Primary Industries and Energy, Australia, Draft of 12.6.1991.

Ball, A.P., Hopkinson, R.B., Farrell, I.D., Hutchison,

J.G.P., Paul, R., Watson, R.D.S., Page, A.J.F., Parker, R.G.F., Edwards, C.W., Snow, M., Scott, D.K., Leone-Ganado, A., Hasting, A., Ghosh, A.C. and Gilbert, R.J.. (1979) - Human botulism caused by Clostridium botulinum type E: The Birmingham outbreak. Quarterly Journal of Medicine, 48, 473-491.

Benelux - draft Code for the Production, Distribution and Sale of Chilled Long-Life Pasteurized Meals, September 1990.

Boden, M. (1991) - New Trends in Vacuum and Gas Flush Packing, European Food and Drink Review, 145-147.

British Meat Manufacturers' Association (BMMA), - Standard for the Production of Bacon and Bacon Joints, BMMA, (BS) 1.4.91.

British Meat Manufacturers' Association (BMMA), - Standard for the Hygienic and Safe Manufacture, Storage and Distribution of Meat and other Food Products, BMMA, (HS) 1.4.91.

British Trout Association, Quality Assurance Code of Practice, (BTA).

Brock, C. - Letter from the Department of Health consulting interested parties on a proposed European Community Council Directive on the Hygiene of Foodstuffs, Department of Health, 20.3.1992.

Brody, A.L. (1990). MAP - a vision of the future. International Conference on MAP. Campden Food and Drink Research Association, Chipping Campden, Glos. UK.

Bryan, F.L. (1992) - Hazard Analysis Critical Control Point Evaluations - A Guide to Identifying Hazards and Assessing Risks Associated with Food Preparation and Storage, World Health Organisation, Geneva.

Campden Food and Drink Research Association - Evaluation of Shelf Life for Chilled Foods, CFDRA Technical Manual No.28 (2nd Edition), CFDRA Shelf Life Working Party, CFDRA, July 1991.

Campden Food and Drink Research Association. - Guidelines for Microbiological Challenge Testing, CFDRA Technical Manual No.20, CFDRA Microbiology Panel Challenge Testing Working Party, CFDRA, December 1987.

Campden Food and Drink Research Association. - Guidelines to the Types of Food Products Stabilised by Pasteurisation Treatments, CFDRA Technical Manual No.27, Part I, CFDRA, April 1992.

Campden Food and Drink Research Association. - Recommendations for the Design of Pasteurisation Processes, CFDRA Technical Manual No.27, Part II, CFDRA, April 1992.

Chai, T.J, & Liang, K.T. (1992) - Thermal Resistance of Spores from Five Type E *Clostridium botulinum* Strains in Eastern Oyster Homogenates, Journal of Food Protection Vol 55(1), 18-22.

Chilled Food Association. - Guidelines for Good Hygienic Practice in the Manufacture, Distribution and Retail Sale of Chilled Foods, CFA, 14.12.89.

Chilled Food Association. - Press Information - New European Chilled Food Body, CFA, 28.10.91.

Chou, J.H., Hwang, P.H. and Malison, M.D. (1988) - An outbreak of type A foodborne botulism in Taiwan due to commercially preserved peanuts. International Journal of Epidemiology, 17, 899-902.

Cockey, R.R. and Tatro M.C. (1974) - Survival studies with spores of *Clostridium botulinum* type E in pasteurised meat of the blue crab, Calinectes sapidus. Appl Microbiol. 27, 629-633

Codex Alimentarius Commission - Recommended International Code of Practice: General Principles of Food Hygiene, 2nd Revision (1985)

Colebatch J. G, Wolff A.H, Gilbert R.J, Mathias C. J, Smith S.E, Hirsch N, Wiles C. M. (1989) - Slow recovery from severe botulism. Lancet, 1216-7.

Critchley E.M.R, Hayes P.J, & Issacs P.E.T. (1989) - Outbreak of botulism in North West England and Wales June 1989. Lancet, ii, 849-853.

Dairy, Food and Environmental Sanitation - Food and Environmental Hazards to Health: Fish Botulism Hawaii 1990, Dairy Food and Environmental Protection, 12, No.3, page 158, March 1992.

Davidson W, D. - Retail store handling conditions for refrigerated foods. Presented at the 80th Annual Convention of the National Food Processors Assn. 1987, Jan 26th, Chicago Ill.

Day, B.P.F. (1990) - A perspective of modified atmosphere packaging of fresh produce in Western Europe, Food Science and Technology Today, Vol 4(4), 215-221.

Day, B.P.F. (1991). Presenting preservation: MAP marches forward, Food Packaging Technology International, 4, 63.

Day, B.P.F. (1991). New atmosphere for international packaging, Food Manufacture International, 7 (3), 28.

Day, B.P.F. (1992). Guidelines for the manufacture and handling of modified atmosphere packaged food products. Campden Food and Drink Research Association Technical Manual 34, CFDRA, Chipping Campden, Glos. UK.

Doyle M. P, (1991) - Evaluating the Potential Risk from Extended - Shelf Life Refrigerated Foods by *Clostridium botulinum* Inoculation Studies, Food Technology, 154-156.

Economist Intelligence Unit Retail Business (Market Reports) No 410, p130 (1992)

European Commission. - Council Directive 89/397/EEC of 14. 6 1989 on the Official Control of Foodstuffs, Official Journal of the European Communities, No. L186, 30.6.89.

European Commission. - Council Directive 91/493/EEC of 22 July 1991 laying down the Health Conditions for the production and the placing on the market of fishery products, Official Journal of the European Communities, No. L268, 24.9.91.

European Commission. - Council Directive 91/497/EEC of 29 July 1991 amending and consolidating Directive 64/433/EEC on health problems affecting intra-Community trade in fresh meat to extend it to the production and marketing of fresh meat, Official Journal of the European Communities, No. L268, 24.9.91.

European Commission. - Council Directive 92/5/EEC of 10 February 1992 amending and updating Directive 77/99/EEC on health problems affecting intra-Community trade in meat products and amending Directive 64/433/EEC, Official Journal of the European Communities, No. L57, 2.3.92.

European Commission. - Draft Commission Decision on a Code of Good Hygienic Practices for the Production of Prepared Meals (Doc VI/2936/90), EC, 1990.

European Commission. - Proposal for a Council Directive on the Hygiene of Foodstuffs (COM(91) 525 final), Official Journal of the European Communities, No. C24, 31.1.92.

European Commission Council Directive 92/46/EEC of 16 June 1992 laying down health rules for the production and placing on the market of raw milk, heat-treated milk and milk based products. Official Journal of the European Communities, No L268, 14.9.92.

Farbar J. M. (1991) - Microbiological Aspects of Modified Atmosphere Packaging Technology - A Review, Journal of Food Protection, Vol 54, No 1, 58-70.

Food Product Intelligence Centre (1992). Chilled food market introductions 1969-1992. Campden Food and Drink Research Association, Chipping Campden. Glos., UK.

French Republic, - Order of 26.6.74 'Regulations for the Hygienic Conditions Concerning Preparation, Preservation, Distribution and Sale of Ready to Eat Meals', Journal Officiel de la Republic Francaise, 16.7.74.

French Republic. - Prolongation of Life Span of Pre-Cooked Food, Modification of Procedures enabling Authorisations to be Obtained, Veterinary Service of Food Hygiene, Service Note DGAL/SVHA/N88/No 8106, 31.5.88.

Gaze, J.E., & Brown G.D. (1990a)- Determination of the growth potential of Clostridium botulinum type E and non-proteolytic type B in "Sous Vide" products at low temperatures, CFDRA Technical Memorandum, September 1990.

Gaze J.E., & Brown G.D, (1990b)- Determination of the heat resistance of a strain of non-proteolytic Clostridium botulinum type B and a strain of type E, heated in Cod and Carrot Homogenate over the temperature range 70 to 90°C. CFDRA Technical Memorandum, October 1990.

Gaze J.E, & Brown G.D, (1991)- Growth of Clostridium botulinum non-proteolytic type B and type E in "Sous Vide" products stored at 2-15°C. CFDRA Technical Memorandum, January 1991.

Gilbert, R.J, and Willis, A.T. (1980) - Botulism. Community Medicine, 2, 25-27.

Gilbert, R.J., Rodhouse, J.C. and Haugh, C.A. 1990. Anaerobes and food poisoning. In: Clinical and Molecular Aspects of Anaerobes; VI Biennial Anaerobe Discussion Group, International Symposium, held at the University of Cambridge, 20-22 July 1989. Wrightson Biomedical Publications Ltd, Petersfield, England.

Graham, A.F. and Lund, B.M. (1991) - The effect of temperature on growth of non-proteolytic type B strains of Clostridium botulinum, Journal of Applied Bacteriology, B71B, pt 6 xxi-xxii.

Grau F. H. & Vanderlinde P.B. (1992) - Occurrence, Numbers, and Growth of Listeria monocytogenes on some Vacuum-Packaged Processed Meats, Journal of Food Protection, Vol 55, 4-7.

Hall, J.D., McCroskey, L.M., Pincomb, B.J. and Hatheway, C.L. (1985). Isolation of an organism resembling Clostridium barati which produces type F botulinal neurotoxin. J. Clin. Microbiol, 21, 654-655.

Harmer, J. 1992 - Room for Improvement (room service menus), Caterer and Hotelkeeper, 19(3), 48-49.

Hauschild, A.H.W. (1989). Clostridium botulinum. In: Foodborne Bacterial Pathogens, ed. M.P. Doyle, pp. 111-189. New York: Marcel Dekker Inc.

Hughes, J.M., Blumenthal J.R., Merson, M.H., Lombard G.L., Dowell, V.R. and Gangarosa, E.J. (1981). Clinical features of types A and B foodborne botulism, Ann.Intern.Med, 95, 442-5.

Illinois State Department of Health, U.S. - Association of Food and Drug Officials - Retail Guidelines - Refrigerated Foods in Reduced Oxygen Packages, Final Draft submitted to Executive Committee,   (undated).

Institute of Food Science and Technology. - Guidelines for the Handling of Chilled Foods 2nd Edition, IFST, 1990.

Jay, J.M. - (1986) Modern Food Microbiology, 3rd edition, London,: Van Nostrand. Reiln hold Company Ltd 1986.

Key Note Publications. - Key Note Report a Market Sector Overview - Chilled Foods, 3rd Edition, Key Note Publications Ltd, 1991.

Lancet, - Foodborne Illness - A Lancet Review, 1991 E. Arnold, London. Leighton, G.R. - Report of the circumstances attending the deaths of eight persons at Loch Maree, Ross-shire. Official Report to the Scottish Board of Health. HMSO: Edinburgh (1923).

Lilley Jnr, T. & Kautter D.A. (1990) - Outgrowth of Naturally Occuring Clostridium botulinum in Vacuum - Packaged Fresh Fish, J. Assoc. of Anal. Chem,73(2), 211-212.

Louis, M.E.St., Peck, S.H.S., Bowering, D., Morgan, G.B., Blatherwick, J., Banerjee, S., Kettyls, G.D.M., Black, W.A., Milling, M.E. Hauschild, A.H.W., Tauxe, R.V. and Blake, P.A. (1988). Botulism from chopped garlic: delayed recognition of a major outbreak. Annals of Internal Medicine, 108, 363-368.

L.S.R.O. - Emerging Issues in Food Safety and Quality for the Next Decade, Center for Food Safety and Applied Nutrition FDA, February 1991,8 , 14-16.

Lund, B. George , S.M. & Franklin, J.G. (1987) - Inhibition of type A and type B (proteolytic) Clostridium botulinum by sorbic acid, Applied and Environmental Microbiology, 53, 935-941.

Lund, B., Graham, A,F, & Brown, D, (1990) - The combined effect of incubation temperature, pH and sorbic acid on the probability of growth of non-proteolytic, type B Clostridium botulinum, J. Appl.bact. 69, 481-492.

Lynt R.K., Kautter D.A. and Solomon, H.M. (1982). Differences and similarities among proteolytic and non-proteolytic strains of Clostridium botulinum types A,B,E and F: A Review. J. Food Prot.45, 466-474.

MacDonald, K.L., Cohen, M.L. and Blake, P.A. (1986). The changing epidemiology of adult botulism in the United States. American Journal of Epidemiology, 124, 794-799.

Mackay-Scollay, E.M. (1958). Two cases of botulism, Journal of Pathology and Bacteriology, 75, 482-485.

Maddox, M. (1991) - The rise and rise of MAP, Food Manufacture, 33-35.

Majewski, C. - Sous-vide - New Technology Catering?, Environmental Health, April 1990, 100-102.

Marketpower Ltd (1986). The packaging of fresh prepared foods - a European survey, Marketpower, London, UK. Marketpower Ltd (1988). European market development in modified atmosphere packaging, Marketpower, London, UK. Maryland (U.S.). - FDA/ Maryland State Health Department Recommendations for Vacuum Packaging of Food in Retail Stores, (undated)

McCroskey, L.M., Hatheway, C.L., Fenicia, L., Pasolini, B. and Aureli, P. (1986). Isolation of an organism which produces type E botulinal toxin that resembles Clostridium butyricum from faeces of an infant with botulism, J. Clin. Microbiol. 23 , 654-655.

McCroskey, L.M., Hatheway, C.L., Woodruff, B.A., Greenberg, J.A. and Jurgonsen P. (1991), Type F botulism due to neurotoxigenic Clostridium barati from an unknown source in an adult. J. Clin. Microbiol. 29, 2618-2620.

Meat and Livestock Commission. - MLC British Meat Catering Service Guide to British Meat, extracts: Hygiene, Handling and Storage, and Equipment, MLC, December 1991, 6-7 & 18-19.

National Food Processors Association (U.S.). - Guidelines for the Developmemnt, Production, Distribution and Handling of Refrigerated Foods, NFPA Microbiology and Food Safety Committee, 1989.

New York State Department of Agriculture and Markets. - Requirements for Vacuum or Modified Atmosphere Packaging at Retail, 27.2.1989.

Notermans, S., Dufrenne, J., Lund B.M. (1990) - Botulism Risk of Refrigerated, Processed Foods of Extended Durability. Journal of Food Protection, Vol 53(12), 1020-24.

Ohye, D.F. & Scott, W.J. (1957) - Studies in the physiology of Clostridium botulinum type E. Australian Journal of Biological Sciences, 10, 85-94.

Okereke, A. & Montville T. J. (1991) - Bacteriocin - Mediated Inhibition of Clostridium botulinum Spores by Lactic Acid Bacteria at Refrigeration and Abuse Temperatures. Applied and Environmental Microbiology, 3423-3428.

O'Mahony, M., Mitchell, E., Gilbert, R.J., Hutchinson, D.N., Begg, N.T., Rodhouse, J.C. and Morris, J.E.(1990). An outbreak of foodborne botulism associated with contaminated hazelnut yoghurt. Epidemiology and Infection, 104, 389-395.

Packaging Strategies (1988). Modified atmosphere packaging: The 'quiet revolution' begins. Packaging Strategies, West Chester, Pennsylvania, USA. Peck, M.W., Fairburn, D.A., and Lund, B.M. (1991) - The effect of lysozyme on growth from heat-damaged spores of non-proteolytic Clostridium botulinum, Journal of Applied Bacteriology, 71, xxii.

Pelletier , A., Hubert, B., Sebald, M. (1991) - Botulism in 1989 and 1990 in France, Bulletin Epidemiologique Hebdomadaire, No 27, p.111.

Powrie, W.D. (1992). New horizons on modified atmosphere packaging for fruits. 9th Foodplas Conference, Plastics Institute of America Inc., Fairfield, New Jersey, USA. Reddy N.R. & Armstrong D.J. and Rhodehamel E. J. & Kautter D.A. (1992) - Shelf-life Extension and Safety Concerns about Fresh Fishery Products Packaged under Modified Atmospheres: A Review, Journal of Food Safety, 12, 87-118.

Rothwell, T.T. (1987). - Flushed with success. Packaging Week, 3 (6),18.

Rothwell, T.T. (1988). A review of current European markets for CA/MAP. Food Eng. 60 (2) 38. Schafheitle, J. M. (1990) - A Basic Guide to Vacuum (Sous-vide) Cooking, Department of Food and Hospitality Management Bournemouth Polytechnic.

Schafheitle, J. M. - The Sous-vide System for Preparing Chilled Meals, British Food Journal, 92(5), 23-27.

Sebald, M. and Saimot, G. (1973). - Circulating toxin, an aid in the diagnosis of type B botulism in man. Ann. Microbiol 124A,61-69.

Segal, M. (1992) - Native Food Preparation Fosters Botulism, FDA Consumer, January-February 1992, 23-26.

Segal, M. (1992) - Botulism in the Entire United States, FDA Consumer, January-February 1992, 27.

Shaw, S., Gabbott, M. and Nowell, L. - The Marketing Activities of British Salmon Smokers 1988, University of Stirling Institute For Retail Studies, Market Report No. 9, November 1989

Schvester, P. (1991). - The effects of noble gases on extending the shelf-life of foods. CAP'91. Schotland Business Research Inc., Princeton, New Jersey. USA. Shone, C. et al, (1985) - Monoclonal Antibody-based Immunoassay for type A Clostridium botulinum toxin is comparable to the mouse bioassay. Appl. Environ. Microbiol, 50, 63-67.

Smith, J. P, et al, (1990) - A Hazard Analysis Critical Control Point Approach (HACCP) to ensure the microbiological safety of sous vide processed meat/pasta product Food Microbiology, 7, 177-198.

Snyder, O.P. - The Application of HACCP Procedures for Sous Vide and Vacuum Packaged Prepared Foods, 1991.

Solomon, H. M., Kautter, D. A., Lynt, R. K. (1982). - Effect of low temperature on growth of non-proteolytic Clostridium botulinum types B and F and proteolytic type G in crab meat and broth. J Food Protect, 45, 516-518.

Sonnabend, O., Sonnnabend, W., Heinzle, R. , Sigrist, T., Dirnhofer, R., Krech, U. (1981) - Isolation of Clostridium botulinum type G and identification of type G botulinal toxin in humans: report of five sudden unexpected deaths. J Infect Dis, 143, 22-27.

Sous Vide Advisory Committee, (1991) - Code of Practice for Sous Vide Catering Systems, SVAC, August 1991.

Stammen. K., Gerdes, G. and Caporaso, F. (1990) - Modified Atmosphere Packaging of Seafood. Food Science and Nutrition Critical Reviews In, 29(5), 301-331.

Stumbo, C. R. (1973) - Thermo Bacteriology in Food Processing,
2nd edition, Academic Press, New York.

Suen, J.C., Hatheway,C.L., Steigerwalt, A.G. and Brenner, D.G., 1988. Genetic confirmation of identities of neurotoxigenic Clostridium barati and Clostridium butmium as agents of infant botulism. Journal of Clinical Microbiology, 26, 2191-2192.

Syndicat National des Fabricants De Plats Préparés (SYNAFAP), France, - Guidelines of Good Hygiene Practices for Prepared Refrigerated Meals, SYNAFAP July 1989.

Szabo, E.A., Pemberton, J.M., snd Desmarchelier, P.M. (1992). Specific detection of Clostridium botulinum type B by using the Polymerase chain reaction. Appl. Environ Microbiol. 58, 418-420.

Tacket, C.O. and Rogawski, M.A. (1989). Botulism. In: Botulinum Neurotoxin and Tetanus Toxin (ed L.L.Simpson), pp.351-378. Academic Press, New York.

U.S. National Advisory Committee on Microbiological Criteria for Foods, - Recommendations for Refrigerated Foods Containing Cooked, Uncured Meat or Poultry Products that are Packaged for Extended Refrigerated Shelf Life and that are Ready-to-Eat or Prepared with little or no Additional Heat Treatment, adopted by the Committee, 31.1.90.

Van Garde, S.J., and Woodburn, M.J. (1987) - Food Discard Practices of Householders. J. Am. Diet. Assn, 87, 322-329.

Walford, D. - Letter from the Department of Health to Environmental Health Departments on the Prevention of Botulism (EL(89)

Walford, D. - Letter from the Department of Health to Environmental Health Departments on the Prevention of Botulism (EL(89) P 145), DH, 29.8.89.

Whiting, R.C. & Naftulin K.A. (1992) - Effect on Headspace Oxygen Concentration on Growth and Toxin Production by Proteolytic Strains of *Clostridium botulinum*, Journal of Food Protection, 55(1), 23-27.

WHO. - Botulism Associated with Rakfisk in Norway, Surveillance Programme for Control of Foodborne Infections and Intoxications in Europe Newsletter, No. 32, April 1992.

SECTION 3: UNPUBLISHED PAPERS

Blood, R.M., Davies, A.R. and Bently, S.D. (1990) - The Microbiological Status of Sous-Vide Products, BFMIRA Leatherhead, Special Project No WO49E, November 1990.

Sheard, M. A. & Rodger, C. (1991) - An investigation of the Heating Efficiency of Equipment Used to Cook Sous Vide Products, Faculty of Cultural and Education Studies - Leeds Polytechnic, April 1991.

GLOSSARY

This glossary is included as an aid to the reader and is not intended to be definitive.

ACIDULANT
Having the ability to increase the acidity

AEROBE
A microorganism that grows at normal atmospheric concentrations of oxygen.

ANAEROBE
A microorganism that is able to grow in the absence of oxygen.

ASEPTIC PACKAGING
Sterilisation of a food followed by filling the food under sterile conditions into pre-sterilised containers which are then sealed.

BACTERIUM
A type of unicellular microorganism that lacks a proper nucleus and reproduces by splitting in two.

CONTROLLED ATMOSPHERE PACKING
The packaging of food in an atmosphere that is different from the normal composition of air, the atmospheric components are precisely adjusted to specific concentrations which are maintained throughout storage.

COOK-CHILL
A food production system based on cooking, rapid cooling and chilled storage.

COOKED
Heated so that all parts of a food or food component have reached a time/temperature equivalent of 70°C for 2 minutes.

CUISINE SOUS VIDE
A system of cooking raw or par-cooked food in a sealed pouch under vacuum. The pouches receive a pasteurisation heat treatment, are cooled and stored under refrigeration at less than 3°C. Products are regenerated using heat, usually by bain marie or microwave, before consumption. Product life is normally 8 to 21 days (some products permitted 42 days under French law).

CURING
A method of food preservation in which meat (particularly pig meat) is permeated with a solution typically containing $NaCl$, and $NaNO_2$ at a temperature of e.g. 4°C

D VALUE
The time required at a given temperature, to reduce the number of viable cells or spores of a given microorganism to 10% of the initial number - usually quoted in minutes.

EC DIRECTIVE
Legislation from the European Community binding upon a Member State as to the result to be achieved within a stated period, but leaving the method of implementation to national governments. (As opposed to an EC Regulation which is directly applicable witout needing confirmation by national Parliaments).

EPIDEMIOLOGY
The study of factors affecting health and disease in populations and the application of this study to the control and prevention of disease.

FACULTATIVE ANAEROBE
A microorganism that is able to grow aerobically or anaerobically.

FOOD POISONING
Any disease of an infectious or toxic nature caused by or thought to be caused by the consumption of food or water.

GRAM
A staining procedure used as an initial step in the identification of bacteria for taxonomic purposes

HACCP
Or Hazard Analysis and Critical Control Points. A structured approach to assesing potential hazards in an operation and deciding which are critical to consumer safety. These critical control points (CCPs), are then monitored in situ and specified remedial action taken if any deviate from safe limits.

HAZARD
Potential to cause harm.

HERMETICALLY SEALED
A closure which constitutes part of a food package and which is designed to prevent microbial contamination of the product.

HOT FILL
Packaging of hot cooked food.

IMMUNOCOMPROMISED
An individual who is unable to mount a normal immune response.

LOW ACID CANNING
Canning of a low acid product with a pH value of above 4.5 and able to support the growth of *Clostridium botulinum* which must be given a minimal thermal process value to reduce the probability of survival of *Clostridium botulinum* spores to be less than 1 in 10^{12} containers. Often termed a 'botulinum cook'.

MICROAEROBE
A microorganism that grows at lower concentrations of oxygen than aerobes.

MICROFLORA
The totality of microorganisms normally associated with a given environment or location.

MODIFIED ATMOSPHERE PACKAGING
The modification of the atmosphere surrounding the food such that usually by the addition of CO_2 the atmosphere is different from the normal composition of air. Product life is extended by refrigerated storage.

NITRITE
A salt used in curing which under appropriate conditions inhibits the growth of vegetative bacteria, also enhances the flavour and colour of cured meats.

PASTEURISATION
A form of heat treatment that kills certain vegetative pathogens and spoilage bacteria in milk and other foods. Temperatures below 100°C are used.

PATHOGEN
Any disease causing microorganism.

pH
An index used as a measure of acidity or alkalinity.

PRESERVATIVE
Any chemical used to prevent or delay microbial and other forms of food spoilage.

PROTEOLYTIC

Having the ability to break down proteins.

PSYCHROTROPH

An organism which can grow at low temperatures (eg 0-5°C) but which has an optimum growth temperature >15°C and an upper limit for growth <30°C.

REDOX POTENTIAL

Oxidation-reduction potential, a measure of a given system to act as a reducing agent, or an oxidizing agent.

RISK

An estimation of the likely occurrence of the hazard (see Hazard)

SHELF-LIFE

The period of time from manufacture to consumption that a food product remains safe under recommended production and storage conditions.

SPOILAGE ORGANISM

A microorganism causing changes in the appearance, taste or smell of food that are generally regarded to be organoleptically unacceptable.

VACUUM PACKING

The removal of all or most of the air within a package without deliberate replacement with another gas mixture.

Z VALUE

The change in temperature centigrade degrees, required for a 10-fold change in the D Value.

Printed in the United Kingdom for HMSO

Dd. 0295538 C20 12/92